Understanding Structural Engineering

From Theory to Practice

Understanding Structural Engineering

From Theory to Practice

Wai-Fah Chen
Salah El-Din E. El-Metwally

CRC Press
Taylor & Francis Group
Boca Raton   London   New York

CRC Press is an imprint of the
Taylor & Francis Group, an **informa** business

CRC Press
Taylor & Francis Group
6000 Broken Sound Parkway NW, Suite 300
Boca Raton, FL 33487-2742

© 2011 by Taylor and Francis Group, LLC
CRC Press is an imprint of Taylor & Francis Group, an Informa business

No claim to original U.S. Government works

Printed in the United States of America on acid-free paper
10 9 8 7 6 5 4 3 2 1

International Standard Book Number: 978-1-4398-2710-9 (Hardback)

Library of Congress Cataloging-in-Publication Data

Chen, Wai-Fah, 1936-
    Understanding structural engineering : from theory to practice / by Wai-Fah Chen and Salah El-Din E. El-Metwally.
       p. cm.
    Summary: "From science to engineering and from theory to practice, it illustrates different breakthroughs, traced back to their origin and placed into prospective. First, the text presents the fundamental laws of mechanics, the theory of elasticity, and the development of the generalized stress-generalized strain concept. Next, it details the era of plasticity. The finite element method comes as an offspring of the generalized stress generalized strain concept. Finally, the authors explore the era of computer simulation to offer a glimpse into the future"--Provided by publisher.
    Includes bibliographical references and index.
    ISBN 978-1-4398-2710-9 (hardback)
    1. Structural engineering. I. El-Metwally, Salah El-Din E. II. Title.

TA633.C26 2011
624.1--dc22
                                            2010048441

Visit the Taylor & Francis Web site at
http://www.taylorandfrancis.com

and the CRC Press Web site at
http://www.crcpress.com

Contents

Preface

This book grew out of the keynote lecture the senior author, Wai-Fah Chen, delivered at the *11th East Asia-Pacific Conference on Structural Engineering and Construction (EASEC-11)* from November 19 to 21, 2008, in Taipei, Taiwan. The full-length paper based on this lecture entitled "Seeing the Big Picture in Structural Engineering" was subsequently published in the *Proceedings of the Institution of Civil Engineers*, May 2009, United Kingdom. The book, as in the paper, sets out to provide "the big picture" guide to the major advances in structural engineering design that have taken place over the last seven decades.

In structural engineering, elasticity and plasticity, mechanics of materials, and continuum mechanics are studied, employed, idealized, simplified, and implemented into engineering practice. The magnitude of difference between the actual performance of a real structure in the real world and the performance predicted on the basis of this drastically simplified theory can only be ascertained by long-term experience and observation, as realistically reflected in building codes supplemented with a variety of safety factors to account for differences.

In this book, we focus on the theories that have stood the test of time and have been widely used in the actual design of structural-engineering solutions. We do not, however, cover historical feats or provide detailed analysis on the design process. Rather, we focus on the way structural engineers deal with ideal material models, ideal structural elements, and systems, and how they apply these simplifications to the formulation of the basic equations of equilibrium and compatibility of a real structural system thereby achieving successful design solutions. We have deliberately omitted any discussion on the theory of structural dynamics, because a palatable treatment of structural dynamics theory in connection with its application to earthquake-engineering designs cannot be accommodated within the space and scope of this book.

These breakthroughs and success stories in the application of mechanics to the design of engineering structures are covered in chronological order: first the fundamental laws of mechanics and materials, then the theory of elasticity, followed by the development of the generalized stress–generalized strain concept, and, finally, the effect of this concept in making the theory of elasticity more practical, which resulted in the adoption of the *allowable stress design* method in specifications worldwide with undue emphasis on safety based on linear elastic analysis.

Like the theory of elasticity in earlier eras, the theory of plasticity with its drastic idealization and simplification provides another success story of applied mechanics, which leads to the powerful limit analysis with lower- and upper-bound theorems for the determination of the load-carrying capacity of structures through the application of the simple plastic-hinge concept, leading to the adoption of the *plastic design* method in steel as well as the *yield line theory* and *strut-and-tie model* in reinforced concrete design codes worldwide.

In recent years, rapid advances in computer technology have spurred the development of structural calculations. There are several numerical analysis approaches to the estimation of stresses, strains, and displacements, but the finite element method is certainly the most versatile and popular. This method is an offspring of the generalized stress–generalized strain concept, which expresses the elastic or plastic relations in terms of structural elements from which the parts of the structure are composed rather than the material treated as a mathematical point, as defined elegantly in the concept of continuum mechanics. The generalized stress–generalized strain concept connects the conventional strength-of-materials approach to the continuum-mechanics-based theory of plasticity, leading to the modern development of finite element solutions in structural engineering. This has allowed to solve almost any structural engineering problem under any condition. As a result of this success, design specifications around the world are being revolutionized, from allowable stress design, to plastic design, to load-resistance-factor design, to the more recent performance-based design—as exemplified by the new Eurocode, American Institute of Steel Construction, and American Concrete Institute codes.

We are now in an age of unlimited desktop computing. Computer simulation has now joined theory and experimentation as a third path for engineering design and performance evaluation and provides us a glance to the future trend of structural engineering in the new century.

Seeing the big picture will enable structural engineers to deal with future theory with idealizations of idealizations and to make it work in the real world of engineering. To help the reader see the big picture of structural engineering, this book consists of the following features:

- It illustrates the key breakthroughs in concept in structural engineering over the last 70 years in a unified manner.
- It presents the science of structural engineering from basic mechanics of materials, to computing, and to the ultimate process of engineering design.
- It shows how we are gaining ground on implementing theory into engineering practice through idealizations and simplifications.
- It explains that seeing the big picture will enable structural engineers to make a difference in the further advancement of the art in the years to come.
- It indicates the modern and future trends in structural engineering and predicts what is to come.

We would like to acknowledge the efforts of Dr. H. M. Nada and Dr. M. M. Gad in preparing the finite element solutions for the problems in Chapters 2 and 4. We are grateful to the sincere efforts of A. A. Khorshed, engineer, who prepared most of the drawings in this book.

Wai-Fah Chen
Salah El-Din E. El-Metwally

Authors

Dr. Wai-Fah Chen has been a well-respected leader in the field of plasticity, structural stability, and structural steel design over the past half century. He has made major contributions to introduce the mathematical theory of plasticity to civil engineering practice, especially in the application of limit analysis methods to the geotechnical engineering field. Having headed the engineering departments at the University of Hawaii and Purdue University, Chen is a widely cited author and the recipient of several national engineering awards, including the 1990 Shortridge Hardesty Award from the American Society of Civil Engineers and the 2003 Lifetime Achievement Award from the American Institute of Steel Construction.

Dr. Chen earned his PhD at Brown University. He is a member of the U.S. National Academy of Engineering and a member of Taiwan's National Academy of Science (Academia Sinica), an honorary member of the American Society of Civil Engineers, a distinguished former professor of civil engineering at Purdue University, and dean of engineering at the University of Hawaii, where he is currently a research professor of civil engineering.

Dr. Salah El-Din E. El-Metwally has vast experience in the structural design of many educational and industrial structures, bridges, and large-scale roof structures. He has been active in research in different areas of structural engineering such as stability, behavior of concrete structures, conceptual design, application of numerical methods in structural engineering, as well as solar energy utilization for about three decades. For many years, he has been an active member in the standing committee and subcommittees of the Egyptian Code for the Design and Construction of Concrete Structures. His academic positions in Egypt include head of the structural engineering department at El-Mansoura University and Tanta University. He was a fellow of Alexander von Humboldt at the Institute of Structural Design, University of Stuttgart, and a visiting professor at the University of Hawaii at Manoa.

Dr. El-Metwally earned his PhD at Purdue University and his MSc at George Washington University. His honors include the State Prize in Structural Engineering and the Encouraging State Prize in Engineering Science, both granted by the Egyptian Academy of Science and Technology, and the Award of Distinction of First Class, granted by the president of Egypt. He is currently a professor of concrete structures at El-Mansoura University.

1 From Science to Engineering

1.1 HISTORICAL SKETCH

The master builders, the designers, and the constructors of the Gothic cathedrals of the Middle Ages used intuition and experience to develop design rules based on simple force equilibrium and treated the material as rigid. This solution process provided the equivalent of what is now known as the lower-bound theorem of plastic limit analysis. This theorem was not proved until more than 500 years later. Modern lower-bound theorem shows that these design rules are safe. These simple design rules have existed from the earliest times for building Greek temples, Roman aqueducts and arch bridges, and domes and vaults. However, tests on real structures showed that the stresses calculated by designers with these rules could not actually be measured in practice.

Galileo, in the seventeenth century, was the first to introduce recognizably modern science into the calculation of structures; he determined the breaking strength of beams but he was way ahead of his time in engineering application. In the eighteenth century, engineers moved away from his proposed "ultimate load" approach, and, until early in the nineteenth century, a formal philosophy of design had been established: a structure should remain elastic, with a safety factor on stress built into the analysis. It was an era of great advancement and a milestone in structural design but one that placed too much emphasis on the undue safety concern based on elastic response under working loads.

Galileo Galilei (1564–1642) was the first to use mathematics in order to describe the law of nature, which is based on observation from his experiments. Isaac Newton (1642–1727) discovered the basic laws of physics in terms of equilibrium condition (or equation of equilibrium) and equation of motion. Robert Hooke (1635–1703) described, in mathematical form, the material response to stress, which he observed in tests. He stated the linear relationship between stress and strain (Hooke's law or constitutive law) as a function of a material constant (elasticity modulus or Young's modulus).

Material continuity without discontinuities or cracks is a logical assumption in solid mechanics. This assumption leads to a mathematical description of geometric relations of a continuous medium known as *continuum* expressed in the form now known as compatibility conditions. For a continuum, the conditions of equilibrium (physics), constitutive (materials), and continuity (geometry) furnish the three sets of basic equations necessary for solutions in any solid mechanics problem in which structural engineering is one of its applications. In short, the mechanics analysis of a given

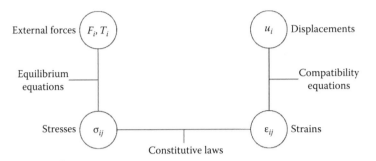

FIGURE 1.1   Interrelationship of the three sets of basic field equations.

structural problem or a proposed structural design must involve the mathematical formulation of the following three sets of equations and solutions:

- Equilibrium equations or motion reflecting laws of physics (e.g., Newton's laws)
- Constitutive equations or stress–strain relations reflecting material behavior (experiments)
- Compatibility equations or kinematical relations reflecting the geometry or continuity of materials (logic)

The interrelationship of these three sets of basic equations is shown in Figure 1.1 for the case of static analysis.

1.2   THE FUNDAMENTALS OF STRUCTURAL ANALYSIS

The basic step of structural analysis is the application of these three sets of equations of equilibrium, compatibility, and constitutive laws. They are the fundamentals for all methods of structural analysis. In general, the three sets of basic conditions are expressed in terms of 15 equations: 3 equilibrium, 6 compatibility, and 6 constitutive equations. The solution of these 15 simultaneous equations should provide solutions of 6 stresses, 6 strains, and 3 displacements at a point in the structure system under consideration. It is the role of mathematics to achieve the solutions of these equations (Sokolnikoff, 1956).

In principle, the solution of the 15 equations for 15 unknowns is possible from a mathematical point of view. However, for real-world applications, a structural engineer must operate with ideal material models and ideal structural systems to reduce drastically the 15 unknowns. The theories of reinforced concrete design, for example, do not deal with real reinforced concrete. They operate with an ideal composite material consisting of concrete and steel, the design properties of which have been approximated from those of real reinforced concrete by a process of drastic idealization and simplification. The same process of simplification and idealization also applies to the formulation of the basic equations of equilibrium and compatibility of a real structural system. This process of simplification and idealization is described in more detail in the chapters that follow.

The first three chapters that follow—The Era of Elasticity (Chapter 2), The Era of Plasticity (Chapter 3), and The Era of Finite Element (Chapter 4)—highlight several key breakthroughs in concepts and simplifications in each era to reach the present state of analysis that is now familiar to structural engineers. These three chapters are essential to the understanding of the remaining chapters—Strut-and-Tie Model (Chapter 5), Advanced Analysis (Chapter 6), and Model-Based Simulation (Chapter 7)—which can be read independently of one another. The breakthroughs described in the first three chapters include, most notably, the following concepts and theorems:

- The generalized stress-generalized strain concept connects the conventional strength-of-materials approach to a continuum-mechanics-based theory of elasticity and plasticity, leading to the modern development of finite element (FE) solutions in structural engineering (Chapter 4).
- The proof of the limit theorems of perfect plasticity provides rational principles for preliminary structural design via simple equilibrium or kinematical processes consistent with engineers' intuitive approaches to design, leading to the modern development of strut-and-tie models (STMs) for structural design in reinforced concrete in particular (Chapter 5).
- The simple plastic-hinge concept enables the direct application of simple plastic theory to steel-frame design in particular, leading to the modern development of advanced analysis for structural design in steel (Chapter 6).

Computer simulation has now combined theory and experimentation as a third path for engineering design and performance evaluation. Simulation is computing, theory is modeling, and experimentation is validation. The major challenges for future structural engineers are the integration and simplification of materials science, structural engineering, and computation and then making them work and applicable for the real world of engineering. The emerging areas of model-based simulation (MBS) in structural engineering are described briefly in Chapter 7.

1.3 ELASTIC ANALYSIS AS A START

To simplify the field equations for a realistic engineering solution, it is more convenient to formulate the elastic or plastic relations in terms of elements from which the parts of the structure are composed rather than the material treated as a mathematical point as defined elegantly in the concept of continuum mechanics. For example, for a structural member such as a beam in a building framework, the basic element or segment can be obtained by cutting through the entire thickness of the beam section. Thanks to this approach, it is then possible to replace the six stress components acting on the cross section of the element by one dominant normal stress resultant—the bending moment, M (generalized stress). Similarly, the corresponding six deformational components can be reduced to one dominant strain resultant—the angle of relative rotation or curvature, φ (generalized strain).

This concept of using the generalized stresses and generalized strains for inelastic structural analysis and design was employed for the first time in 1952 by Prager

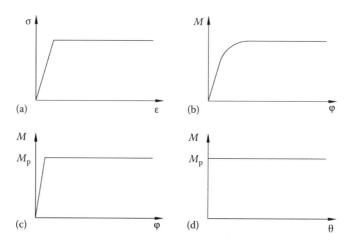

FIGURE 1.2 Development of plastic-hinge concept: (a) elastic–perfectly plastic stress–strain relation, (b) moment–curvature relation, (c) idealized moment–curvature relation, and (d) idealized moment–rotation relation.

in establishing his general theory of limit design, and, later in 1959, it was utilized prominently by Hodge in his popular text on the plastic analysis of structures (Hodge, 1959; Prager, 1952). It took great insight to fully understand the impact by unifying the conventional strength-of-materials approach and the modern theory of plasticity and limit design in a consistent manner.

The relationship between the value of the bending moment M and the angle of relative rotation φ for the ends of the section represents the material behavior of that structural element (generalized stress–generalized strain relation, Figure 1.2b). The relationship is linear and reversible for a linear elastic material, as observed by Hooke, before yielding or crack or under working load condition. With this simplification, it has become possible to develop solutions for structural members and frames. These solutions so obtained are called *strength-of-material* solutions. Thanks to this simplification, the complex local stress and strain states in a real sizable element of a real structure are avoided and the field of application of the theory of elasticity, described in Chapter 2, and that of plasticity, described in Chapter 3, can be broadened significantly. This expansion and generalization resulted in the development of modern structural theories, among them several structural elements including bar elements, plate elements, shell elements, and FEs (Chapter 4).

This study and mathematical formulation of engineering structures have led to a formal three-stage process in mechanics operation, which can be summarized as follows:

- First, the relationships between stresses in a structural element and the generalized stresses acting on the surface of the element are determined by using equilibrium equations.
- Second, the relationships between deformations of the material in the element and the generalized strains on the surface of the element are

established through a kinematical assumption such as "plane section before bending remains plane after bending."

- Finally, the generalized stress–generalized strain relations are derived through the use of stress–strain relations of the material.

This mechanics operation becomes particularly clear when this process is applied to a particular structural element and material model under consideration. Chapter 2 shows the application for a linear elastic material along with its many engineering solutions suitable for use in allowable stress design code in earlier years, while Chapter 3 shows its companion application for a plastic material along with its further development and simplification leading to plastic-hinge concept with the subsequent development of simple plastic theory used widely in plastic design code in steel in later years.

Chapter 4 shows this formulation process along with how it can be developed naturally for modern FE analysis. Since the solution for the simultaneous FE equation must be carried out numerically in an incremental form by computer, many numerical procedures were developed in the period from 1970 to the 1980s including the necessity of developing an efficient iterative process to deal with load-path dependency of the inelastic material behavior. As a result of this progress, together with the rapid advancement in computer power, large amount of numerical data were generated in a variety of structural engineering applications during this period. With this large amount of database generated, the probability theory was utilized to analyze these results leading to the development of reliability-based code in recent years. It was an era of great advancement.

1.4 PLASTIC ANALYSIS AS A FURTHER PROGRESS

The idealization of elastic–perfectly plastic behavior of material beyond the elastic range opened the door to a new era of mechanics. Introducing this idealization in the formulation of the generalized stress–generalized strain relation led to several advanced relations of structure elements. For instance, the elastic–perfectly plastic uniaxial stress–strain relation in Figure 1.2a leads to the generalized stress–generalized strain relation (moment–curvature relation) of cross section shown in Figure 1.2b. This moment–curvature relationship must be further idealized in order to develop simple plastic theory for engineering practice. This leads, for example, to ignoring strain-hardening and also to eliminating entirely the effect of time from the calculations. This further idealization is illustrated in Figure 1.2c, leading to the concept of *plastic hinge*, Figure 1.2d, by ignoring further the relatively small elastic strains near collapse of a structure. This further idealization of perfect plasticity to deal with the complex plastic behavior of the structural element gives powerful limit theorems of plasticity (Drucker et al., 1952), which made it possible to estimate the collapse load of a variety of structure systems including beams, plates, and shells in a direct manner.

The upper- and lower-bound theorems of limit analysis of perfect plasticity provide an excellent guide for preliminary design as well as for analysis of structures.

- *Lower-bound theorem.* If an equilibrium distribution of moment can be found, which balances the applied loads, and is everywhere below the plastic moment or at the plastic moment value, the structure will not collapse or will be just at the point of collapse.
- *Upper-bound theorem.* The structure will collapse if there is a compatible pattern of plastic failure mechanism for which the rate at which the external forces work equals or exceeds the rate of internal dissipation.

The lower-bound theorem states that the structure will adjust itself to carry the applied load if at all possible. It gives lower-bound or safe values of the collapse loading. The maximum lower bound is the collapse load itself. The upper-bound theorem states that if a plastic failure mechanism exists, the structure will not stand up. It gives upper-bound or unsafe values of the collapse loading. The minimum upper bound is the collapse load itself.

Historically, engineers in the past, based on intuition, developed many solutions for weak-tension material (based on equilibrium only) and for ductile materials (based on kinematics only), which have now been justified by the rigorous theorems of limit analysis. The theorems of limit analysis thus represent a very powerful tool nowadays to estimate the collapse load of structures or structural members without having to go through a very tedious calculation procedure. Further discussions on the development and applications of the theory of plasticity and limit analysis to structural design are given in more detail in Chapter 3.

In the case of lower-bound solution of limit analysis, only equilibrium and yield criterion are satisfied; equilibrium is satisfied for stress or generalized stress. The crude solution so obtained represents a good and quick guidance for the structural engineer. It can be used to verify some refined solutions from other methods. The lower-bound method is especially useful for application to tension-weak material, for example, stones or concrete. Hence, the safety of monumental structures such as cathedral can be checked very well with such a method.

In the case of upper-bound solution only kinematics and yield criterion are satisfied. The method is very powerful for ductile materials and even applicable to materials with limited ductility but with some modification to the solution procedure. The quick estimate of the collapse load of a structure is of great value, not only as a simple check for a more refined computer analysis but also as a basis for preliminary engineering design. The method, for example, can be used to make a quick check to verify solutions obtained from some sophisticated FE analysis in particular.

The structural applications of the limit theorems started with the development of the simple plastic theory for steel building design (Neal, 1957) and were extended to the development of yield line theory for reinforced concrete slab design (Nielsen, 1964). Limit theorems have been explored carefully for applications to stability problems in soil mechanics (Chen, 2007), complemented by applications to the metal-forming process (Johnson, 1986) and studied thoroughly in metal-matrix-composites applications (Dvorak and Bahei-El-Din, 1982), among others.

1.5 FINITE ELEMENT ANALYSIS AS A LOGICAL EXTENSION

The development of the FE analysis was a logical extension of the mechanics analysis involving mathematical formulation of the three sets of basic equations and solutions as described previously. First, the concept of generalized stress and generalized strain allowed dealing with a FE instead of a material point in a structure. Second, the principle of virtual work was utilized for the formulation of equilibrium instead of force balance, which simplified the solution process significantly. Third, by assuming an appropriate shape function of an element, compatibility between strains in the element and its nodal displacements was conveniently justified. These simplifications made it possible to obtain engineering solutions of almost any structure of any geometry and of any material model.

The FE method with powerful computers enabled engineers to implement realistic geometry and accurate material models into the analysis. Hence, it has become possible to obtain not only the collapse load of a structure but also the deformations under any loading level including even the post-peak behavior. As a result, it has become possible to apply the theory of stability with the theory of plasticity to simulate the actual behavior of structural members and frames with great confidence. It was the first time we were able to replace the costly full-scale tests with computer simulation. As a result of such progress, together with a rapid advancement in computing power, large amounts of numerical data were generated in a variety of structural engineering applications during this era.

The following is a brief summary of the kind of numerical data that were generated through the FE analysis for structural members and frames in the 1970s. As a result of these data, the limit-state approach to design was advanced and new specifications in steel design were issued in the 1980s. More complete description of these advancements is given in Chapter 4.

1970s—Numerical studies of member-strength equations:

- Beam strength equation leading to beam design curve
- Column strength equation leading to column design curve
- Beam-column-strength equation leading to beam-column interaction design curve
- Biaxially loaded column strength equation for plastic design in steel building frames

These developments were summarized in the two-volume treatise by Chen and Atsuta (2007a,b).

1980s—Limit states to design:

- Development of reliability-based codes
- Publication of 1986 AISC/LRFD specification in the United States (AISC, 1986) and Europe (ECCS, 1984, 1991)
- Introduction of second-order elastic analysis to design codes

- Explicit consideration of semirigid connections in frame design (now known as "partially restrained construction") in the United States (Chen and Kim, 1998) and Europe (ECCS, 1992)

These developments were summarized in the book by Chen and Lui (1992).

1.6　STM AS A POWERFUL TOOL

With the advancement in material modeling and FE idealization, the computational process has not only become more powerful but also more complicated and time consuming. For daily practice it is necessary to rely on simplified analysis but with adequate accuracy. The lower-bound and upper-bound solutions of limit analysis serve this purpose realistically and conveniently. The equilibrium method has been used since ancient times as in the Egyptian pyramids, structures of arched form, and monumental structures. The recent proof of this method supports the ancient engineering practice and helps expand the method to modern applications of reinforced concrete structures.

One of the most important advancements in reinforced concrete in recent years is the extension of lower-bound-limit-theorem-based design procedures to shear, torsion, bearing stresses, and the design of structural discontinuities such as joints and corners. The STM is developed for such a purpose and is based on the lower-bound theorem of limit analysis. In this model, the complex stress distribution in the structure is idealized as a truss carrying the imposed loading through the structure to its supports. Like a real truss, an STM consists of compression struts and tension ties interconnected at nodes. Using the stress legs similar to those shown in Figure 1.3, a lower-bound stress field that satisfies equilibrium and does not violate failure criteria at any point can be constructed easily to provide a safe estimate of load-carrying capacity on the reinforced concrete structures (Chen and Han, 1988).

The STM has been well developed over the last two decades and it was presented in several texts (see for example Schlaich and Schäfer, 1991) as a standard method for shear, joints, and support bearing design. The STM method was also introduced in the AASHTO LRFD Specifications (ASCE, 1998; AASHTO, 1998) as well as in

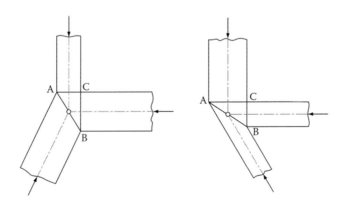

FIGURE 1.3　Using stress legs as truss members to produce a stress field at a stress joint.

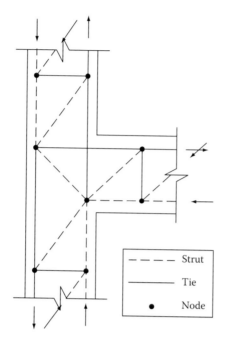

FIGURE 1.4 Application of the STM to reinforced concrete joint design.

the ACI 318 building code (ACI 318-08). A typical example of STM for a common structural joint design is shown in Figure 1.4.

STMs are derived from the flow of forces within structural concrete regions, namely, those of high shear stresses, where Bernoulli hypothesis of flexure, plane sections before bending remain plane after bending, does not apply. Those regions are referred to as discontinuity or disturbance regions (or simply D-regions), in contrast to those regions where Bernoulli hypothesis is valid, and are referred to as Bernoulli or bending regions (or simply B-regions). The flow of forces in D-regions can be traced through the concept of truss, thus named truss model or STM, which is a generalization of the truss model. The concept of STM as a lower-bound solution is illustrated in Chapter 3. More discussions on the applications of the method to a variety of different D-regions in reinforced concrete design are given in Chapter 5.

1.7 ADVANCED ANALYSIS FOR STEEL FRAME DESIGN AS THE CURRENT PROGRESS

In current engineering practice, there is a fundamental two-stage process in the design operation:

- The forces acting on the structural members are determined by conducting an elastic structural system analysis.
- The sizes of various structural members are selected by checking against the ultimate strength equations specified in design codes.

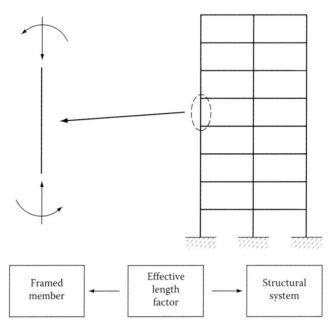

FIGURE 1.5 Interaction between a structural system and its component members using the *K* factor concept.

The interaction behavior between individual members and their structural system is accounted for approximately by the use of the effective length factor *K* concept as illustrated schematically in Figure 1.5. However, despite its popular use in current practice as a basis for design, the effective length approach has the following major limitations:

- It cannot reflect the inelastic distributions of internal forces in a structural system.
- It cannot provide information on the failure mechanisms of a structural system.
- It is not easy to implement in an integrated computer design application.
- It is a time-consuming process by calculating every *K* factor for each separate member capacity check.

Furthermore, some of these difficulties are more so on seismic designs since additional questions are frequently asked:

- How is the structure going to behave during an earthquake?
- Which part of the structure is the most critical area?
- What will happen if part of the structure yields or fails?
- What might happen if forces greater than the code has specified occur?

Considering these limitations and drawbacks and the rapid advancement of computing power, the second-order inelastic analysis approach or the so-called advanced analysis approach provides an alternative approach to structural analysis and design. Nevertheless, this approach consumes tremendous computation efforts, and in order to overcome such a demand the practical advanced analysis has been developed (Chen, 2009a). The practical advanced analysis as presented in Chapter 6 is an elastic-plastic-hinge-based analysis, modified to include the geometry imperfections, gradually yielding and residual stress effects, and semirigid connections. In this approach, all those aforementioned drawbacks associated with using K factor are overcome. There is no need to compute the effective length factor, yet it will produce almost identical member sizes as those of the LRFD method (Chen and Kim, 1997).

1.8 COMPUTER-BASED SIMULATION AS THE FUTURE TREND

We are now in a desktop environment for unlimited computing. Computer simulation has now combined theory and experimentation as a third path to scientific knowledge. Simulation plays an increasing critical role in all areas of science and engineering. Exciting examples of these simulations are occurring in areas such as automotive crashworthiness for component design in auto industry, *Boeing 777* for system design and manufacturing in aerospace, and the next generation Space Telescope (Hubble II) for system design, assembly, and operation in space engineering. One key branch of this new discipline is MBS, whose objective is to develop the capability for realistically simulating the behavior of complex systems under the loading and environmental conditions that the systems may experience during their lifetimes.

Simulation does not replace observation and physical experimentation but complements and enhances their value in the synthesis of analytical models. It provides a framework for combining theory and experimentation with advanced computation. Besides massive numerical computations, high-performance computers permit the use of other tools, such as visualization and global communications using advanced networks, all of which contribute to the ability to understand and control the physical processes governing complex systems.

MBS is based on the integration of mechanics, computing, physics, and materials science for predicting the behavior of complex engineering and natural systems. MBS allows engineers and researchers to investigate the entire life cycle of engineered systems and assists in decisions on the design, construction, and performance in civil and mechanical systems. Reliable and accurate MBS tools will permit the design of engineering systems that cost less and perform better. MBS promises to reduce design cycle times while increasing system life span.

The emerging areas of MBS in structural engineering will notably include the following topics:

1. From the present structural system approach to the life-cycle structural analysis and design covering construction sequence analysis during construction, performance analysis during service, and degradation and deterioration analysis during maintenance, rehabilitation, and demolition

2. From the present FE modeling for continuous media to the finite block types of modeling for tension-weak materials, which will develop cracks and subsequently change the geometry and topology of the structure
3. From the present time-independent elastic and inelastic material modeling to the time-dependent modeling reflecting material degradation and deterioration science

These emerging areas of research and application are inherently interdisciplinary in science and engineering, where computation plays the key role. Scientists provide a consistent theory for application, and structural engineers must continue to face the reality of dealing with idealizations of idealizations of these theories in order to make them work and applicable to the real world of engineering. A brief description of the MBS to structural engineering applications is presented in Chapter 7.

1.9 SUMMARY

Over the last few decades, remarkable developments have occurred in computer hardware and software. Advancement in computer technology has spurred the development of structural calculations ranging from the simple strength-of-materials approach in early years, to the FE type of structural analysis for design in recent years, and to the modern development of scientific simulation and visualization for structural problems in the years to come.

Table 1.1 summarizes briefly the "major advances" of structural engineering that can be attributed to the "breakthroughs" of mechanics formulation, material modeling, or computing power where new knowledge has been implemented in structural engineering and, in some measure, the structural engineering practice has been fundamentally changed. These "success stories" fall into one of the following three categories: *mechanics*, *materials*, and *computing* as tabulated briefly in Table 1.1 (Chen, 2009b).

A topic on which significant progress has been made in recent years is the determination of the load-carrying capacity of structures through the application of the theory of plasticity. This is in contrast to the earlier era design with undue emphasis on linear elastic analysis. Engineering specifications contained rules that help engineers avoid most of the errors of overdesign or under-design with guidelines derived from experience and tests. However, rules based on past experience work well only for designs lying within the scope of that range. They cannot be relied on outside of that range. Ideally, the design guidelines and rules should be derived from sound physical and mathematical principles.

Similar to the theory of elasticity in earlier eras, the theory of plasticity in later years provides one of these success stories of applied mechanics that leads to the development of modern design guidelines and rules. The mathematical theory of plasticity enables us to go beyond the elastic range in a time-independent but theoretically consistent way for inelastic structural analysis and design.

The introduction of the concept of generalized stresses and generalized strains for structural elements and the establishment of the general theory of limit analysis and design in the 1950s laid the foundation for the revolution in structural engineering

TABLE 1.1

The Interaction of Mechanics, Materials, and Computing and the Advancement of Structural Engineering Practice

Mechanics	Materials	Computing	Structural Analysis and Design
Strength-of-materials formulation: closed form solutions by series expansion, numerical solutions by finite difference	Linear elasticity	Slide rule and calculator environment	Strength-of-materials approach to structural engineering in the early years: • Allowable stress design with K factor • Amplification factor for second-order effect • Moment distribution or slope deflection methods for load distribution in framed structures • Member by member design process • Design rules based on allowable strength of members from tests with built-in safety factors
Limit analysis methods: mechanism method and equilibrium method, plastic-hinge concept	Perfect plasticity	Slide rule and calculator environment	Simple plastic analysis method for steel frame design in the early years: • Plastic analysis and design with K factor • Amplification factor for second-order effects • Upper- and lower-bound methods for frame design • Member by member design process • Design rules based on ultimate strength of members from tests
FE formulation using shape function and virtual work equation: generalized stresses and generalized strains concept	General plasticity	Mainframe computing environment	FE approach to structural engineering in recent years: • Development of member-strength equations with probability and reliability theory • Development of reliability-based codes • Limit states to design with K factor • Direct calculation of second-order effect • Member by member design approach • Design rules based on load factor and resistance factor concept by mathematical theory

(*continued*)

TABLE 1.1 (continued)

The Interaction of Mechanics, Materials, and Computing and the Advancement of Structural Engineering Practice

Mechanics	Materials	Computing	Structural Analysis and Design
Advanced analysis: combining theory of stability with theory of plasticity	General plasticity	Desktop computing with object-oriented programming	Second-order inelastic analysis for direct frame design as the current progress: • Structural system approach to design without K factor and amplification factor • Explicit consideration of the influence of structural joints in the analysis/design process • Development of performance-based codes • Consideration of "structural fuse" concept in design • Design based on maximum strength of the structural system without having to carry out member by member strength check
MBS based on the integration of mechanics, computing, physics, and materials science	Deterioration science or aging	High-performance computing	Large-scale simulation of structural system over its life-cycle performance analysis: • Numerical challenges: proper modeling of discontinuity and fracture or crack for tension-weak materials • Software challenges: radically different scales in time and/or space • Material challenges: from time-independent elastic and inelastic material model to time-dependent modeling reflecting material degradation and deterioration science • Design process includes modeling (physics), simulation (computing), visualization (software), and verification (experiment)

in subsequent years. The adoption of plastic analysis methods in steel specifications started the revolution in the 1960s. Thanks to the rapid advancement of computing power beginning in the 1970s, the study of mechanics and mathematical formulation subsequently focused on the study of structural elements from which the parts of the structure are composed rather than the material itself. Thanks to this approach, the field of application on the theory of plasticity to structural engineering has broadened appreciably.

In the more recent years, various analysis approaches to the estimation of stress, strain, and displacement including analytical, numerical, physical, and analog techniques have advanced and are readily available to the engineering profession. In particular, the FE technique is most versatile and popular. As a result of this success, design specifications around the world have been undergoing several stages of revolutionary changes from the allowable stress design, to plastic design, to load-resistance factor design, and to the more recent performance-based design.

We are now in a desktop environment for unlimited computing. Computer *simulation* has now combined the *theory* and *experimentation* as a third path for engineering design and performance evaluation. Simulation is computing, theory is modeling, and experimentation is validation of the results. As a structural engineer, we must continue to face the reality of dealing with idealizations of idealizations of these science-based theories in order to make them work and applicable to the real world of engineering. Seeing the big picture of our past achievements in structural engineering will enable us to make a difference in the further advancement of structural engineering in the years to come. This is described in this book.

REFERENCES

AASHTO, 1998, *AASHTO LRFD Bridge Specifications*, 2nd edn., American Association of State Highway and Transportation Officials, Washington, DC.

American Institute of Steel Construction, 1986 (2005), *Load and Resistance Factor Design Specification for Structural Steel Buildings*, AISC, Chicago, IL.

ASCE-ACI Committee 445 on Shear and Torsion, 1998, Recent approaches to shear design of structural concrete, *Journal of Structural Engineering, ASCE*, 124, 12, 1375–1417.

Chen, W. F., 2007, *Limit Analysis and Soil Plasticity*, Elsevier, Amsterdam, the Netherlands, 1975, Reprinted by J. Ross, Orlando, FL.

Chen W. F., 2009a, Toward practical advanced analysis for steel frame design, *Journal of the International Association for Bridge and Structural Engineering (IABSE), SEI*, 19, 3, 234–239.

Chen, W. F., 2009b, Seeing the big picture in structural engineering, *Proceedings of the Institution of Civil Engineers*, 162, CE2, 87–95.

Chen, W. F. and Atsuta, T., 2007a, *Theory of Beam-Columns, Vol. 1, In-Plane Behavior and Design*, McGraw-Hill, New York, 1976, Reprinted by J. Ross, Orlando, FL.

Chen, W. F. and Atsuta, T., 2007b, *Theory of Beam-Columns, Vol. 2, Space Behavior and Design*, McGraw-Hill, New York, 1977, Reprinted by J. Ross, Orlando, FL.

Chen, W. F. and Han, D. J., 1988, *Plasticity for Structural Engineers*, Springer-Verlag, New York.

Chen, W. F. and Kim, S. E., 1977, *LRFD Design Using Advanced Analysis*, CRC Press, Boca Raton, FL.

Chen, W. F. and Kim, Y. S., 1998, *Practical Analysis for Partially Restrained Frame Design*, Structural Stability Research Council, Lehigh University, Bethlehem, PA.

Chen, W. F. and Lui, E. M., 1992, *Stability Design of Steel Frames*, CRC Press, Boca Raton, FL.

Drucker, D. C., Prager, W., and Greenberg, H., 1952, Extended limit design theorems for continuous media, *Quarterly Applied Mathematics*, 9, 381–389.

Dvorak, G. J. and Bahei-El-Din, Y. A., 1982, Plasticity analysis of fibrous composites, *Journal of Applied Mechanics*, 49, 327–335.

ECCS, 1984, Ultimate limit state calculation of sway frames with rigid joints, ECCS-Publication No. 33, Brussels, Belgium.

ECCS, 1991, *Essential of Eurocode 3 Design Manual for Steel Structures in Buildings*, ECCS-Advisory Committee 5, No. 65, Brussels, Belgium.

ECCS, 1992, Analysis and design of steel frames with semi-rigid joints, ECCS-Publication No. 61, Brussels, Belgium.

Hodge, P. G., 1959, *The Plastic Analysis of Structures*, McGraw-Hill, New York.

Johnson, W., 1986, The mechanics of metal working plasticity, *in Applied Mechanics Update*, C.R. Steele and G.S. Springer, eds., ASME, New York.

Neal, B. G., 1957, *The Plastic Methods of Structural Analysis*, Chapman & Hall, London, U.K.

Nielsen, M. P., 1964, Limit analysis of reinforced concrete slabs, *Acta Polytechnica Scandinavica*, Civil Engineering Building Construction Service, No. 26.

Prager, W., 1952 (1955), Sectional address, *Proceedings of the Eighth International Congress on Theoretical and Applied Mechanics*, Istanbul, Turkey, August 20–28.

Schlaich, J. and Schafer, K., 1991, Design and detailing of structural concrete using strut-and-tie models, *The Structural Engineer*, 69, 6, 113–125.

Sokolnikoff, I. S., 1956, *Mathematical Theory of Elasticity*, 2nd edn., McGraw-Hill, New York.

2 The Era of Elasticity

2.1 FUNDAMENTALS OF ELASTICITY

2.1.1 BASIC FIELD EQUATIONS

In order to establish a solution in continuum mechanics, three basic sets of relations have to be fulfilled: (1) equilibrium conditions, which guarantee that the body is always in equilibrium; (2) compatibility conditions, which guarantee that the body remains continuous; and (3) constitutive relations, which connect stresses and strains of a material behavior. These relations can be expressed in tensor notations as follows.

2.1.1.1 Equilibrium

The equilibrium conditions for an arbitrary volume, V, Figure 2.1, are

$$\int_V (\sigma_{ij,j} + F_i)\,dV = 0 \tag{2.1}$$

where
  σ_{ij} is the stress tensor
  F_i is the body force

Equation 2.1 may be written in the usual $(X, Y, \text{and } Z)$ notation as

$$\frac{\partial \sigma_x}{\partial x} + \frac{\partial \tau_{xy}}{\partial y} + \frac{\partial \tau_{xz}}{\partial z} + F_x = 0$$

$$\frac{\partial \tau_{yx}}{\partial x} + \frac{\partial \sigma_y}{\partial y} + \frac{\partial \tau_{yz}}{\partial z} + F_y = 0 \tag{2.2}$$

$$\frac{\partial \tau_{zx}}{\partial x} + \frac{\partial \tau_{zy}}{\partial y} + \frac{\partial \sigma_z}{\partial z} + F_z = 0$$

where
  σ_x, σ_y, and σ_z represent the normal stress components
  τ_{xy}, τ_{yz}, … represent the shear stress components

2.1.1.2 Compatibility

The compatibility conditions (or conditions of body continuity) can be expressed as

$$\varepsilon_{ij,kl} + \varepsilon_{kl,ij} - \varepsilon_{ik,jl} - \varepsilon_{jl,ik} = 0 \tag{2.3}$$

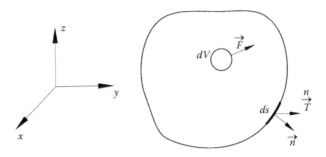

FIGURE 2.1   Equilibrium of a material body.

Upon expanding these expressions, the following can be obtained with respect to the usual (X, Y, and Z) notation:

$$\frac{\partial^2 \varepsilon_x}{\partial y^2} + \frac{\partial^2 \varepsilon_y}{\partial x^2} = 2\frac{\partial^2 \varepsilon_{xy}}{\partial x \partial y}$$

$$\frac{\partial^2 \varepsilon_y}{\partial z^2} + \frac{\partial^2 \varepsilon_z}{\partial y^2} = 2\frac{\partial^2 \varepsilon_{yz}}{\partial y \partial z}$$

$$\frac{\partial^2 \varepsilon_z}{\partial x^2} + \frac{\partial^2 \varepsilon_x}{\partial z^2} = 2\frac{\partial^2 \varepsilon_{zx}}{\partial z \partial x}$$

$$\frac{\partial}{\partial x}\left(-\frac{\partial \varepsilon_{yz}}{\partial x} + \frac{\partial \varepsilon_{zx}}{\partial y} + \frac{\partial \varepsilon_{xy}}{\partial z} \right) = \frac{\partial^2 \varepsilon_x}{\partial y \partial z}$$

$$\frac{\partial}{\partial y}\left(-\frac{\partial \varepsilon_{zx}}{\partial y} + \frac{\partial \varepsilon_{xy}}{\partial z} + \frac{\partial \varepsilon_{yz}}{\partial x} \right) = \frac{\partial^2 \varepsilon_y}{\partial z \partial x}$$

$$\frac{\partial}{\partial z}\left(-\frac{\partial \varepsilon_{xy}}{\partial z} + \frac{\partial \varepsilon_{yz}}{\partial x} + \frac{\partial \varepsilon_{zx}}{\partial y} \right) = \frac{\partial^2 \varepsilon_z}{\partial x \partial y}$$

$$(2.4)$$

where

ε_x, ε_y, and ε_z represent the normal strain components

ε_{xy}, ε_{yz}, ... represent the shear strain components

For a material element in volume, V, the three equations of equilibrium, Equation 2.2 and the six equations of compatibility, Equation 2.4, give a sum of nine equations. On the other hand, the total number of unknowns is 15 (6 stress components, 6 strain components, and 3 displacement components). The remaining six equations necessary to obtain a solution are the material-dependent equations (constitutive relations). The interrelations between the three sets of relations are schematically illustrated in Figure 2.2.

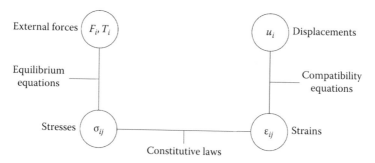

FIGURE 2.2 Interrelations between mechanics variables.

2.1.1.3 Constitutive Relations

The constitutive relations can be expressed in tensor notation as follows:

$$\sigma_{ij} = C_{ijkl}\varepsilon_{kl} \tag{2.5}$$

where C_{ijkl} is the material elastic constant tensor. Equation 2.5 is a simple generalization of Hooke's law experiment in a simple tension test, and, therefore, it is referred to as the generalized Hooke's law. This equation can be written in matrix form for an isotropic linear elastic material as follows:

$$\{\sigma\} = [C]\{\varepsilon\} \tag{2.6}$$

where the matrix $[C]$ is called the elastic constitutive or elastic moduli matrix and is given by

$$[C] = \frac{E}{(1+v)(1-2v)}\begin{bmatrix} (1-v) & v & v & 0 & 0 & 0 \\ v & (1-v) & v & 0 & 0 & 0 \\ v & v & (1-v) & 0 & 0 & 0 \\ 0 & 0 & 0 & \frac{(1-2v)}{2} & 0 & 0 \\ 0 & 0 & 0 & 0 & \frac{(1-2v)}{2} & 0 \\ 0 & 0 & 0 & 0 & 0 & \frac{(1-2v)}{2} \end{bmatrix} \tag{2.7}$$

where
  E is Young's modulus
  v is Poisson's ratio

2.1.2 SOLUTION PROCESS: AN ILLUSTRATION

Linear elasticity is based on two fundamental assumptions: the stress–strain relation is linear and is reversible. The first assumption allows for the application of

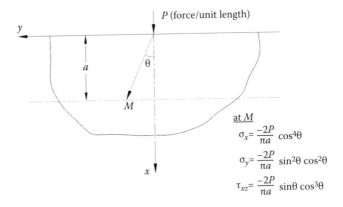

at M

$$\sigma_x = \frac{-2P}{\pi a} \cos^4\theta$$

$$\sigma_y = \frac{-2P}{\pi a} \sin^2\theta \cos^2\theta$$

$$\tau_{xz} = \frac{-2P}{\pi a} \sin\theta \cos^3\theta$$

FIGURE 2.3　Half-space problem.

the principle of superposition. The second assumption means that the material is load-path independent, which permits for total load application, that is, no need to go to incremental solution or follow load history. This leads to a simple and direct formulation of the equations of mechanics at a point in a body. Nevertheless, the solution of the 15 equations, except in very special cases, represents a great mathematical challenge, and in some cases is impossible to achieve without exploring nontraditional thinking. Figure 2.3 shows the solution of the problem of half-space under concentrated load, one of the few problems that can be handled from a direct solution of the 15 equations.

2.2　THE CONCEPT OF GENERALIZED STRESS AND GENERALIZED STRAIN

2.2.1　Introduction

In the concept of continuum mechanics, the field relations derived at a material point can be used to establish solutions for limited applications. In order to widen the scope of applications, these field relations should be formulated in terms of elements from which the structure is composed, for example, bar, plate, shell, and finite element (FE). For example, for a bar element instead of the six stress components, σ_{ij}, at a point one can deal with the bending moment (generalized stress), M. Correspondingly, the six strain components, ε_{ij}, are replaced with the angle of relative rotation or curvature (generalized strain), φ. The same concept applies to other structure elements such as plate, shell, FE, etc.

One may ask how to decide on which variable a generalized stress or generalized strain is. Bending moment in a beam can be considered as a generalized stress, whereas other variables, such as shear, cannot be considered as generalized stresses because the deformations associated with their correspondent strains are negligible. In a beam, for example, the curvature signifies the dominant deformation (bending deformation), and therefore it is considered as a generalized strain and its correspondent stress variable (moment) is considered as a generalized stress.

Another question may arise about the significance of this concept and its impact. The answer is very simple; for example, in a beam problem with continuum mechanics approach, the solution searches for 15 variables (15 equations in 15 unknowns). On the other hand, in a beam analogy (bending theory) the solution searches for one variable. The reduction in the effort is thus very obvious when using this concept since the effort is an exponential function of the number of variables. In this regard, Einstein was asked once why he would use one type of soap instead of two as other people do. He replied: two soaps are too many variables, one is enough and much simpler.

In the following sections, equilibrium is utilized to develop the generalized stress in terms of the stress tensor. Compatibility is utilized to derive the generalized strain in terms of the strain tensor. The constitutive relations are employed to connect the generalized stresses and the generalized strains. These developments are given for different structure elements, for example, bar, plate bending, shell element, and FE.

2.2.2 Bar Element as a Start

2.2.2.1 Axially Loaded Element

For a bar element subjected to an axial force (generalized stress), P, causing a uniform axial stress, σ, equilibrium leads to

$$P = \sigma A \tag{2.8}$$

where A is the cross-sectional area of the bar. The stress, σ, causes an axial strain, ε, which is associated with a change in bar length (generalized strain), Δ. Based on compatibility the generalized strain is related to the strain as follows:

$$\varepsilon = \frac{\partial \Delta}{\partial x} \tag{2.9}$$

Upon integration of the previous equation

$$\Delta = \varepsilon L \tag{2.10}$$

where L is the bar length. The generalized stress, P, can be related to the generalized strain, Δ, upon employing Hooke's law:

$$\sigma = E\varepsilon \tag{2.11}$$

where E is Young's modulus. This leads to

$$P = \sigma A = E\varepsilon A = E\frac{\Delta}{L}A$$

or

$$P = \left(\frac{EA}{L}\right)\Delta \tag{2.12}$$

The term (EA/L) connecting the two variables P and Δ is the bar axial stiffness.

2.2.2.2 Flexural Element

The same simple and systematic steps followed for a bar under axial force can be followed for the development of the variables of a bar under pure bending. Consider a bar element of length ΔS subjected to end moments, M, such that the element deforms as shown in Figure 2.4. Upon applying equilibrium at the end section of the element

$$M = \int \sigma y \, dA \tag{2.13}$$

The generalized strain (or curvature), φ, can be related to the strain in a beam, ε, upon the adoption of linear strain distribution within the section depth according to Bernoulli's hypothesis (plane sections perpendicular to the neutral axis before bending remain plane and perpendicular to the neutral axis after bending). Thus,

$$\varepsilon = \varphi y \tag{2.14}$$

The generalized stress, M, is related to the generalized strain, φ, through the application of the stress–strain relation $\sigma = E\varepsilon$:

$$M = \int \sigma y \, dA = \int E\varepsilon y \, dA = \int E\varphi y^2 \, dA = E\varphi \int y^2 \, dA$$

or

$$M = EI\varphi \tag{2.15}$$

where I is the moment of inertia of the beam cross section. It should be noted that only the bending moment can be considered as a generalized stress, whereas other

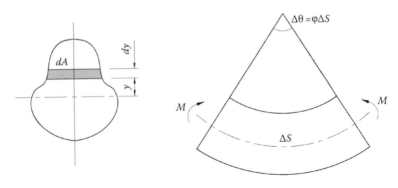

FIGURE 2.4   Bar element under pure bending.

parameters such as shear or torsion cannot be, since the deformations associated with their correspondent stresses are negligible.

2.2.3 Plate Element as a Next Step

2.2.3.1 Compatibility

For a plate element in Cartesian coordinates, the thickness is denoted as h and the deflection as w. Since there is no normal force applied to the end sections of the plate segment shown in Figure 2.5, the neutral surface is assumed to coincide with the middle surface of the plate. Based on Kirchhoff's hypothesis for thin plates with small deflection (plane sections perpendicular to the neutral surface before bending remain plane and perpendicular to the neutral surface after bending), the elongation of a fiber parallel to the x- or y-axis is proportional to its distance z from the middle surface. The generalized strains (curvature of the deflection in the x- and y- directions, φ_x and φ_y, respectively) can be taken as

$$\varphi_x = \frac{1}{r_x} = -\frac{\partial^2 w}{\partial x^2} \tag{2.16a}$$

$$\varphi_y = \frac{1}{r_y} = -\frac{\partial^2 w}{\partial y^2} \tag{2.16b}$$

where r_x and r_y are the radii of curvatures in the x- and y-directions, respectively. From kinematics, the strains at a distance z from the middle surface in the x- and y-directions, ε_x and ε_y, respectively, are

$$\varepsilon_x = \frac{\partial u}{\partial x} = \frac{\partial}{\partial x}\left(-z\frac{\partial w}{\partial x}\right) = -z\frac{\partial^2 w}{\partial x^2} \tag{2.17a}$$

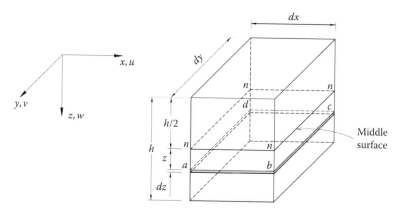

FIGURE 2.5 Plate bending element.

$$\varepsilon_y = \frac{\partial v}{\partial y} = \frac{\partial}{\partial y}\left(-z\frac{\partial w}{\partial y}\right) = -z\frac{\partial^2 w}{\partial y^2} \tag{2.17b}$$

where u and v are the displacement components in the x- and y-directions, respectively.

2.2.3.2 Constitutive Relations

From the generalized Hooke's law

$$\varepsilon_x = \frac{\sigma_x}{E} - v\frac{\sigma_y}{E} \tag{2.18a}$$

$$\varepsilon_y = -v\frac{\sigma_x}{E} + \frac{\sigma_y}{E} \tag{2.18b}$$

where v is Poisson's ratio. This leads to

$$\sigma_x = \frac{E}{1-v^2}(\varepsilon_x + v\varepsilon_y) \tag{2.19a}$$

$$\sigma_y = \frac{E}{1-v^2}(v\varepsilon_x + \varepsilon_y) \tag{2.19b}$$

Upon substitution of Equation 2.17 into Equation 2.19

$$\sigma_x = \frac{-Ez}{1-v^2}\left(\frac{\partial^2 w}{\partial x^2} + v\frac{\partial^2 w}{\partial y^2}\right) \tag{2.20a}$$

$$\sigma_y = \frac{-Ez}{1-v^2}\left(v\frac{\partial^2 w}{\partial x^2} + \frac{\partial^2 w}{\partial y^2}\right) \tag{2.20b}$$

2.2.3.3 Equilibrium

From equilibrium, the generalized stresses M_x and M_y are related to the stresses σ_x and σ_y as follows:

$$M_x = \int_{-h/2}^{h/2} \sigma_x z\, dz \tag{2.21a}$$

$$M_y = \int_{-h/2}^{h/2} \sigma_y z\, dz \tag{2.21b}$$

In this regard, the twisting moments, M_{xy} and M_{yx}, and shearing forces, Q_x and Q_y, cannot be considered as generalized stresses since the deformations associated with their correspondent stresses are negligible. Upon substituting Equation 2.20 into Equation 2.21 and integrating

$$M_x = -D\left(\frac{\partial^2 w}{\partial x^2} + v\frac{\partial^2 w}{\partial y^2}\right) \qquad (2.22a)$$

$$M_y = -D\left(v\frac{\partial^2 w}{\partial x^2} + \frac{\partial^2 w}{\partial y^2}\right) \qquad (2.22b)$$

where

$$D = \frac{Eh^3}{12(1-v^2)} \qquad (2.23)$$

The generalized stresses, M_x and M_y, can be related to the generalized strains (curvatures), φ_x and φ_y, by solving Equations 2.16 and 2.22:

$$M_x = D(\varphi_x + v\varphi_y) = D\left(\frac{1}{r_x} + \frac{v}{r_y}\right) \qquad (2.24a)$$

$$M_y = D(v\varphi_x + \varphi_y) = D\left(\frac{v}{r_x} + \frac{1}{r_y}\right) \qquad (2.24b)$$

2.2.4 SHELL ELEMENT AS A FURTHER EXTENSION

The most common shell structures are thin cylindrical shells under axisymmetric loading, as in water tanks, towers, etc. Therefore, they are taken as an example of shell structures to show how generalized stresses and generalized strains are derived and related.

For this element, Figure 2.6, there are three generalized strains and three corresponding generalized stresses: (1) the radial strain (which is the tangential strain of the middle surface), ε_θ, and its corresponding tangential force, N_θ; (2) the curvature of the cylinder wall in the x-direction, φ_x, and the corresponding moment, M_x; and (3) the curvature of the cylinder wall in the tangential direction, φ_θ, and the corresponding moment, M_θ.

2.2.4.1 Equilibrium

Upon considering the equilibrium of the shell element in Figure 2.6 in the x- and y-directions and the moment about the y-axis (and neglecting higher-order terms), the following relations can be obtained:

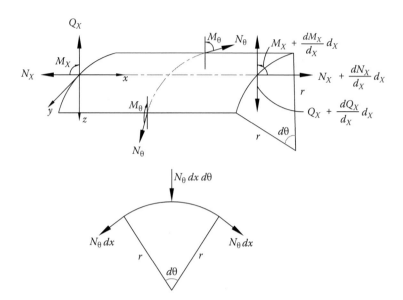

FIGURE 2.6   Element of a circular cylindrical shell.

$$\left(\frac{dN_x}{dx}\,dx\right)r\,d\theta = 0 \tag{2.25}$$

$$\frac{dQ_x}{dx}+\frac{N_\theta}{r}=p \tag{2.26}$$

$$\frac{dM_x}{dx}-Q_x=0 \tag{2.27}$$

Equation 2.25 means that N_x should be equal to either a constant value or zero; in this derivation, it is considered equal to zero. Upon differentiating Equation 2.27 with respect to x and substituting from Equation 2.26, the following can be obtained:

$$\frac{d^2M_x}{dx^2}+\frac{N_\theta}{r}=p \tag{2.28}$$

2.2.4.2   Kinematics for Membrane Action

The kinematical relations and constitutive relations are discussed for membrane action first and later for bending behavior. Define the membrane strains in the x- and radial directions as ε_x and ε_θ, respectively, and the displacements in the x- and radial directions as u and w, respectively (Figure 2.7). Then,

$$\varepsilon_x=\frac{du}{dx} \tag{2.29}$$

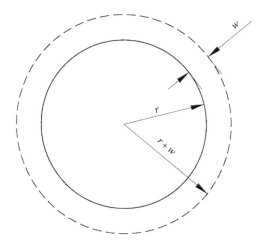

FIGURE 2.7 Radial deformation of a circular cylindrical shell element.

$$\varepsilon_\theta = \frac{w}{r} \tag{2.30}$$

2.2.4.3 Constitutive Relations

From the generalized Hooke's law,

$$\varepsilon_x = \frac{\sigma_x}{E} - v\frac{\sigma_\theta}{E} = \frac{N_x}{Et} - v\frac{N_\theta}{Et} = \frac{1}{Et}(N_x - vN_\theta) \tag{2.31}$$

$$\varepsilon_\theta = -v\frac{\sigma_x}{E} + \frac{\sigma_\theta}{E} = -v\frac{N_x}{Et} + \frac{N_\theta}{Et} = \frac{1}{Et}(-vN_x + N_\theta) \tag{2.32}$$

From Equations 2.31 and 2.32, the following values of N_x and N_θ in terms of ε_x and ε_θ can be obtained:

$$N_x = \frac{Et}{1-v^2}(\varepsilon_x + v\varepsilon_\theta) \tag{2.33}$$

$$N_\theta = \frac{Et}{1-v^2}(v\varepsilon_x + \varepsilon_\theta) \tag{2.34}$$

For an open cylinder, $N_x = 0$, that is,

$$\varepsilon_x = -v\varepsilon_\theta \tag{2.35}$$

Then,

$$N_\theta = Et\varepsilon_\theta \tag{2.36}$$

2.2.4.4 Kinematics for Bending Action

For bending strain, define the strains in the x- and radial directions as ε_x^b and ε_θ^b, and their corresponding stresses as σ_x^b and σ_θ^b, respectively. Based on Kirchhoff hypothesis the strains ε_x^b and ε_θ^b are linear functions of the distance z from the neutral axis, Figure 2.8; that is,

$$\varepsilon_x^b = z\varphi_x \tag{2.37}$$

The notation φ_x, Figure 2.8, is used here for the curvature of the cylinder in the x-direction; however, other notations such as k_x may be commonly used in text books:

$$\varphi_x = \frac{\partial^2 w}{\partial x^2} \tag{2.38}$$

In order to assess ε_θ^b refer to Figure 2.8,

$$\varepsilon_\theta + \varepsilon_\theta^b = \frac{(r-z+w)\,d\theta - (r-z)\,d\theta}{(r-z)\,d\theta} \approx \frac{w}{r}\left(1+\frac{z}{r}\right) \approx \varepsilon_\theta\left(1+\frac{z}{r}\right) \tag{2.39}$$

Since $(z/r) \ll 1$, it can be neglected in the above equation; hence,

$$\varepsilon_\theta^b \approx 0 \tag{2.40}$$

This will also lead to

$$\varphi_\theta = \frac{\varepsilon_\theta + \varepsilon_\theta^b}{r} = \frac{\varepsilon_\theta}{r} = \frac{w}{r^2} \approx 0 \tag{2.41}$$

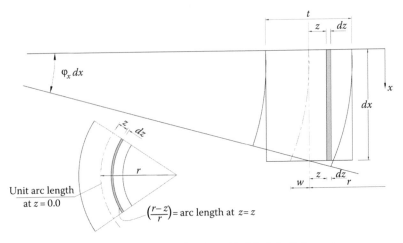

FIGURE 2.8 Bending deformation of a circular cylindrical shell element.

2.2.4.5 Constitutive Relations

From the generalized Hooke's law,

$$\sigma_x^b = \frac{E}{1-v^2}\left(\varepsilon_x^b + v\varepsilon_\theta^b\right) = \frac{E}{1-v^2}(z\varphi_x) \tag{2.42}$$

2.2.4.6 Equilibrium

Thus, the generalized stress M_x is

$$M_x = \int_{-t/2}^{t/2} \sigma_x^b z\, dz = \frac{E}{1-v^2} \int_{-t/2}^{t/2} \varphi_x z^2 dz = \frac{Et^3}{12(1-v^2)}\varphi_x = D\varphi_x \tag{2.43}$$

The stress σ_θ^b can be obtained from Hooke's law as follows:

$$\sigma_\theta^b = \frac{E}{1-v^2}\left(v\varepsilon_x^b + \varepsilon_\theta^b\right) = \frac{E}{1-v^2}\left(v\varepsilon_x^b\right) = \frac{E}{1-v^2}(vz\varphi_x) = \frac{vE}{1-v^2}(z\varphi_x) \tag{2.44}$$

Hence, the moment M_θ is

$$M_\theta = \int_{-t/2}^{t/2} \sigma_\theta^b z\, dz = \frac{vE}{1-v^2} \int_{-t/2}^{t/2} \varphi_x z^2 dz = \frac{vEt^3}{12(1-v^2)}\varphi_x = vD\varphi_x \tag{2.45}$$

In summary, the generalized stresses, N_θ and M_x, and the generalized strains, ε_θ and φ_x, are related as follows:

$$N_\theta = Et\varepsilon_\theta \tag{2.46}$$

$$M_x = D\varphi_x \tag{2.47}$$

The curvature $\varphi_\theta \approx 0$; therefore, the moment $M_\theta = vM_x$ cannot be considered a generalized stress.

2.2.5 FINITE ELEMENT AS A RECENT PROGRESS

2.2.5.1 Kinematics

In the FE method, Figure 2.9, formulation starts with the kinematical (compatibility) conditions. In this step, the generic displacements of an element (internal displacements within an element), $\{u\}$, are related to the generalized strains (the nodal displacements of the element), $\{q\}$, by means of assumed shape functions, $[N]$. This assumption is equivalent to Bernoulli's assumption in beams under bending

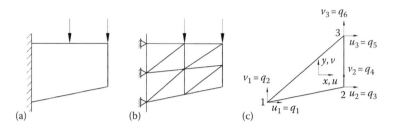

FIGURE 2.9 Sample finite element: (a) structural element; (b) discretization; and (c) finite element.

(plane sections perpendicular to the neutral axis before bending remain plane and perpendicular to the neutral axis after bending) or Kirchhoff's hypothesis in plates under bending:

$$\{u\} = [N]\{q\} \tag{2.48}$$

With the displacement within the element, the strain vector, $\{\varepsilon\}$, at any point within the element can be obtained by differentiation of $\{u\}$ in Equation 2.48

$$\{\varepsilon\} = [B]\{q\} \tag{2.49}$$

where $[B]$ is composed of the derivatives of the shape function, $[N]$.

2.2.5.2 Equilibrium

The next step is to impose the equilibrium condition in order to obtain the relation between the generalized stresses (nodal force vector), $\{F\}$, and the internal stress vector at any point, $\{\sigma\}$. This step can be achieved upon the application of the principle of virtual work. This demonstrates the fact that the principle of virtual work is nothing other than a principle of equilibrium. Thus,

$$\{F\} = \int [B]^t \{\sigma\} \, dV \tag{2.50}$$

2.2.5.3 Constitutive Relations

The third step is to impose the constitutive relations that can be expressed in the following general form:

$$\{\sigma\} = [C]\{\varepsilon\} \tag{2.51}$$

where $[C]$ is called the elastic constitutive or elastic moduli matrix. Upon substitution of (2.49) and (2.51) into (2.50)

$$\{F\} = [k]\{q\} \tag{2.52}$$

where

$$[k] = \int [B]^t [C][B] dV \qquad (2.53)$$

Thus, the formulation of generalized stresses and generalized strains and their relation in the FE is the same as other structure elements such as bars or plates, except in the application of equilibrium and compatibility. Equilibrium is justified through the principle of virtual displacement and compatibility is achieved through the assumed shape function.

2.3 THEORY OF STRUCTURES

2.3.1 GENERALIZED STRESS–GENERALIZED STRAIN RELATIONS OF A BEAM MEMBER

In the preceding section it was illustrated how the generalized stress (moment), M, and generalized strain (curvature), φ, of a beam element are related. This relation is utilized to develop the relation between the generalized stress (end moment), M, and the generalized strain (end rotation), θ, of a beam member.

2.3.1.1 No Sway

Figure 2.10a illustrates a beam member AB with two end moments (generalized stresses), M_A and M_B, associated with end rotations (generalized strains), θ_A and θ_B. The moment at a section located at a distance x from the origin of coordinates, $M(x)$, can be written as

$$M(x) = -M_A + \left(\frac{M_A + M_B}{L} \right) x \qquad (2.54)$$

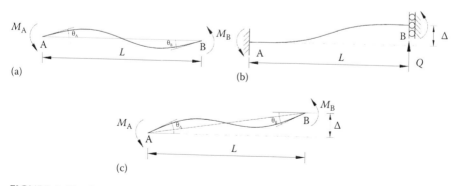

FIGURE 2.10 Beam member subjected to end moments and sway: (a) end moments; (b) sway; and (c) end moments and sway.

From the generalized stress–generalized strain relation of a beam element, the curvature at a section x, $\varphi(x)$, is

$$\varphi(x) = \frac{M(x)}{EI} = \frac{1}{EI}\left\{\left(\frac{M_A + M_B}{L}\right)x - M_A\right\} \tag{2.55}$$

The curvature, $\varphi(x)$, is related to the deflection, $v(x)$, and hence the moment as follows:

$$\frac{d^2 v(x)}{dx^2} = \varphi(x) = \frac{1}{EI}\left\{\left(\frac{M_A + M_B}{L}\right)x - M_A\right\} \tag{2.56}$$

Upon integrating Equation 2.56 twice, and introducing the boundary conditions, $v|_{x=0}=0$ and $v|_{x=L}=0$, the integration constants can be determined. Then, the equation of deflection will be

$$v(x) = \frac{1}{EI}\left\{\left(\frac{M_A + M_B}{6L}\right)x^3 - \frac{M_A}{2}x^2 + \left(\frac{2M_A - M_B}{6}\right)Lx\right\} \tag{2.57}$$

The equation of rotation, $\theta(x)$, will be

$$\theta(x) = \frac{1}{EI}\left\{\left(\frac{M_A + M_B}{2L}\right)x^2 - M_A x + \left(\frac{2M_A - M_B}{6}\right)L\right\} \tag{2.58}$$

Applying Equation 2.58 at member ends A and B

$$\theta_A = \frac{L}{6EI}(2M_A - M_B) \tag{2.59a}$$

and

$$\theta_B = \frac{L}{6EI}(-M_A + 2M_B) \tag{2.59b}$$

Solving Equations 2.59a and b in terms of M_A and M_B leads to

$$M_A = \frac{EI}{L}(4\theta_A + 2\theta_B) \tag{2.60a}$$

$$M_B = \frac{EI}{L}(4\theta_B + 2\theta_A) \tag{2.60b}$$

Equations 2.60a and b represent the generalized stress–generalized strain relation of a beam member with no sway.

2.3.1.2 Sway

If the beam member experiences sway only, Figure 2.10b, the generalized stress will be the shear, Q, and the corresponding generalized strain will be the vertical displacement (sway), Δ. For this case, it can be easily proven, as derived for the case of no sway, that

$$Q = \frac{12EI}{L^3}\Delta \tag{2.61}$$

The shear, Q, will be associated with two end moments, $M_A = M_B = -QL/2$; hence,

$$M_A = M_B = \frac{-6EI}{L^2}\Delta \tag{2.62}$$

The relation in Equation 2.62 is advantageous in limiting the number of variables to one generalized stress. Thus, Equations 2.60a and b and 2.62 can be combined to give the following generalized stress–generalized strain relations for the case of a beam member with sway (Figure 2.10c):

$$M_A = \frac{EI}{L}\left(4\theta_A + 2\theta_B - 6\frac{\Delta}{L}\right) \tag{2.63a}$$

$$M_B = \frac{EI}{L}\left(4\theta_B + 2\theta_A - 6\frac{\Delta}{L}\right) \tag{2.63b}$$

Equations 2.60a and b or 2.63a and b are commonly known as the *slope-deflection equations*.

The concept of generalized stresses and generalized strains has been widely used along with the principle of superposition to develop many successful methods of structural analysis such as the slope-deflection method and the moment distribution method. In addition, it has been used to solve many mechanics problems such as column buckling. Such a contribution is illustrated in the following sections.

2.3.2 SINGLE SPAN BEAM PROBLEMS

The solution of single span beam problems can be obtained from the direct application of the slope-deflection equations and with the enforcement of the appropriate boundary conditions. Examples of these problems are given in Figure 2.11, along with their solutions. For the problem in Figure 2.11c, the end moment at A is set

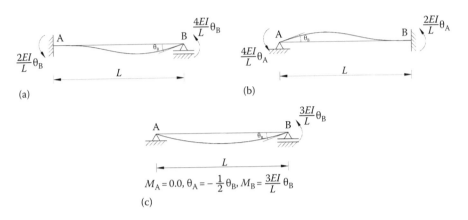

FIGURE 2.11   Examples of single span beam problems: (a) carryover to fixed end A, (b) carryover to fixed end B, and (c) carryover to hinged end A.

equal to zero in Equation 2.60a leading to $\theta_A = -\theta_B/2$, which upon substitution in Equation 2.60b gives the illustrated result. The moment at the fixed end of the beams in Figure 2.11a and b is the carryover moment, which is half the moment applied at the other end (the end free to rotate), that is, the carryover factor in this case is equal to 0.5. On the other hand, in the beam of Figure 2.11c the carryover moment and hence the carryover factor are equal to zero.

For the simply supported beam subjected to uniform load in Figure 2.12a, the end rotation can be obtained by following the same solution procedure in the previous section for obtaining the slope-deflection equations. The solution of the beam subjected to two end moments in Figure 2.12b can be obtained from the slope-deflection equations. Upon using the principle of superposition, the solutions of the previous two beams can be used to obtain the fixed end moments of the beam in Figure 2.12c through the enforcement of the boundary conditions. The same procedure can be followed to obtain the fixed end moments for other cases.

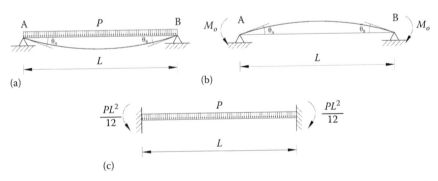

FIGURE 2.12   Fixed end moments: (a) uniform load, (b) equal end moments, and (c) using the principle of superposition.

2.3.3 SLOPE-DEFLECTION METHOD FOR SIMPLE STRUCTURES

The slope-deflection method is one of the successful methods used for the analysis of statically indeterminate skeletal structures and for other applications. It is a form of the stiffness method where the unknowns are the displacements of the joints, that is, the size of the problem depends on the number of unspecified degrees of freedom (degree of kinematic indeterminacy). The application of the method is illustrated in the following example.

The continuous beam shown in Figure 2.13 has two degrees of freedom (the rotations at B and C), which means that the solution leads to the formation of two simultaneous linear equations in θ_B and θ_C. The solution starts with writing the slope-deflection equations of each span as an isolated beam. In these equations, we write the moment at each joint of the span, which is the summation of the fixed end moment due to applied loads and the moment associated with the end rotations or sway (or settlement). Then, we apply equilibrium at joints B and C (summation of end moments at each joint is equal to the external moment applied at the joint, which is zero in this case). Thus, two simultaneous linear equations in θ_B and θ_C are derived, and, hence, these joint rotations are obtained. Consequently, the end moments can be calculated followed by the end shear.

The same procedure of slope-deflection method used in beams applies to frames except in those with sway where equilibrium equation(s) is written for the sway. For example, for the frame in Figure 2.14 the number of degrees of freedom is 3 (the rotations at B and C and the lateral sway, Δ_B or Δ_C). The sway equation is obtained from the equilibrium of the forces in the direction of the sway; in this case, the external horizontal loads and the shear at A and D. The shear is obtained from member equilibrium.

In this method, it is obvious that the equations of the individual spans are developed from the three basic conditions, equilibrium, compatibility, and constitutive relations, as illustrated earlier. Also, in the complementary steps of the solution, compatibility is the initiation of the solution, the rotation at any joint is controlled, and the formation of the slope-deflection equations is obtained by applying the same three sets of conditions. In addition, equilibrium is enforced in order to obtain the simultaneous linear equations of the unknown rotations. Nevertheless, as a result of adopting the generalized stress and generalized strain concept, the solution is much easier here in comparison with the continuum mechanics approach.

2.3.4 MOMENT DISTRIBUTION METHOD FOR FRAME STRUCTURES

The moment distribution method is another successful method for the analysis of statically indeterminate beams and frames. It is a form of the displacement method in

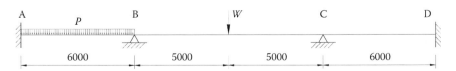

FIGURE 2.13 Example of a continuous beam.

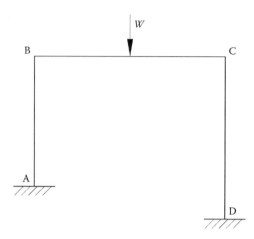

FIGURE 2.14   Example of a frame with a sway.

which equilibrium, compatibility, and constitutive conditions are used to obtain the properties of individual members with the aid of the generalized stress–generalized strain concept. These properties include the fixed end moments, stiffness, and carryover factor, as illustrated in the preceding section. The rigid joints are assumed to be locked and the corresponding moments (fixed end moments) are calculated. Then, the out-of-balance moment at each joint is calculated and distributed between the members connected to the joint according to their stiffness. This step is followed by transferring a percentage of the correction moment between the joints of each member according to the carryover factor. Subsequently, the joints' balance is checked and another adjustment of the moments connected to the joints is achieved followed by the carryover moment. This last iterative procedure is repeated until convergence takes place.

2.3.5 STRUCTURAL DESIGN

It has been illustrated how the three sets of conditions, equilibrium, compatibility, and constitutive conditions, are utilized to develop successful methods for structural analysis. With the introduction of the concept of the generalized stress and generalized strain, the effort of analysis has been remarkably reduced. Two variables replaced the continuum mechanics stress and strain tensors and the constitutive relations with the implementation of Bernoulli's hypothesis. The solution will always search for one unknown, M, and the other unknown comes from the generalized stress–generalized strain relation.

After performing structural analysis and obtaining the internal forces, the stress, σ, can be calculated and checked against the allowable value, σ_{all}:

$$\sigma = \frac{M}{I} y \le \sigma_{all} \tag{2.64}$$

The allowable value is provided by the code and is based on real tests. This value should guarantee safety and in order to do so it accounts for residual stresses and other environmental factors. The code is meant by other issues such as how to avoid local buckling, bracing, etc.

2.4 THEORY OF STRUCTURAL STABILITY

2.4.1 GENERALIZED STRESS–GENERALIZED STRAIN RELATIONS OF A BEAM-COLUMN MEMBER

Similar to the case of beam member, the relation between generalized stress (end moment), M, and generalized strain (end rotation), θ, of a beam-column member can be developed. For the framed member in Figure 2.15a, upon writing the equilibrium relations in the *deformed geometry*, the following relations can be obtained for the case of no sway:

$$M_A = \frac{EI}{L}(s_{ii}\theta_A + s_{ij}\theta_B) \tag{2.65a}$$

$$M_B = \frac{EI}{L}(s_{ji}\theta_A + s_{jj}\theta_B) \tag{2.65b}$$

where the stability functions, s_{ii}, s_{ij}, s_{ji}, and s_{jj}, are as given in Table 2.1. In case there is a sway of value Δ, Figure 2.15b, the generalized stress–generalized strain relations will be

$$M_A = \frac{EI}{L}\left(s_{ii}\theta_A + s_{ij}\theta_B - (s_{ii} + s_{ij})\frac{\Delta}{L}\right) \tag{2.66a}$$

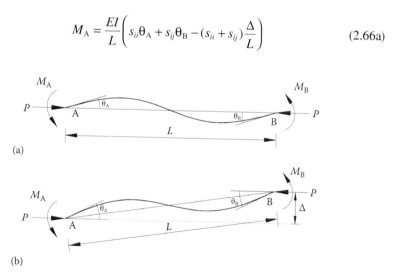

(a)

(b)

FIGURE 2.15 Beam-column member subjected to end moments and sway: (a) no sway and (b) sway.

TABLE 2.1

Stability Functions for a Beam-Column

	Condition of Axial Load, P		
Function	Compression	Tension	Zero
$s_{ii} = s_{jj}$	$\dfrac{kL \sin kL - (kL)^2 \cos kL}{2 - 2\cos kL - kL \sin kL}$	$\dfrac{(kL)^2 \cosh kL - kL \sinh kL}{2 - 2\cosh kL + kL \sinh kL}$	4
$s_{ij} = s_{ji}$	$\dfrac{(kL)^2 - kL \sin kL}{2 - 2\cos kL - kL \sin kL}$	$\dfrac{kL \sinh kL - (kL)^2}{2 - 2\cosh kL + kL \sinh kL}$	2

Note: $k = \sqrt{\dfrac{P}{EI}}$

$$M_B = \frac{EI}{L}\left(s_{ji}\theta_A + s_{jj}\theta_B - (s_{ji} + s_{jj})\frac{\Delta}{L} \right) \tag{2.66b}$$

Equations 2.65a and b or 2.66a and b are commonly referred to as the *slope-deflection equations*.

2.4.2 Buckling Analysis of Structural Members

The generalized stress–generalized strain relations of a beam-column member (or slope-deflection equations) are the analytical tool of a beam-column. This is illustrated by the following three examples.

Example 2.1

This example, Figure 2.16a, describes a beam-column with one fixed end and one hinged end. In this member, the rotation at end A is zero, and, therefore, the slope-deflection equations at ends A and B are

$$M_A = \frac{EI}{L}(s_{ij}\theta_B) \tag{2.67a}$$

$$M_B = \frac{EI}{L}(s_{ij}\theta_B) \tag{2.67b}$$

The moment at joint B, $M_B = 0$; hence, from Equation 2.67b and since $\theta_B \neq 0$

$$s_{jj} = 0$$

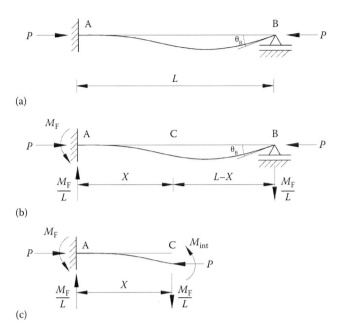

(a)

(b)

(c)

FIGURE 2.16 Beam-column with one fixed end and one hinged end: (a) beam-column with fixed-hinged ends, (b) forces and moments, and (c) a free body.

or

$$\frac{kL \sin kL - (kL)^2 \cos kL}{2 - 2\cos kL - kL \sin kL} = 0 \qquad (2.68)$$

Since $kL \neq 0$, Equation 2.68 is simplified to the following characteristic equation of the beam-column

$$\tan kL - kL = 0 \qquad (2.69)$$

The value of kL that satisfies Equation 2.69 is

$$kL = 4.4934$$

Thus,

$$k^2 L^2 = 20.19 = \frac{PL^2}{EI} \qquad (2.70)$$

that is, the critical load of the beam-column is

$$P_{cr} = \frac{20.19EI}{L^2} = \frac{\pi^2 EI}{(0.7L)^2} \qquad (2.71)$$

The critical load of this member could have been obtained by solving the differential equation of equilibrium. The equilibrium equation of the segment AC, Figure 2.16c, of this member is

$$-M_{int} + Py - M_F\left(1 - \frac{x}{L}\right) = 0 \tag{2.72}$$

Since

$$M_{int} = -EIy'' \tag{2.73}$$

the equilibrium equation can be written as

$$y'' + k^2 y - \frac{M_F}{EI}\left(1 - \frac{x}{L}\right) = 0 \tag{2.74}$$

which has a general solution of the form

$$y(x) = C_1 \sin kx + C_2 \cos kx + \frac{M_F}{P}\left(1 - \frac{x}{L}\right) \tag{2.75}$$

After imposing the boundary conditions, Equation 2.75 would lead to the same characteristic equation of the member, Equation 2.69.

Example 2.2

This example, Figure 2.17, describes a beam-column with one fixed end and one guided end. In this member, the rotation at both ends A and B is zero, and, therefore, the slope-deflection equations at ends A and B are

$$M_A = M_B = \frac{EI}{L}\left[-(s_{ii} + s_{ij})\frac{-\Delta}{L}\right] = \frac{EI\Delta}{L^2}(s_{ii} + s_{ij}) \tag{2.76}$$

From the member equilibrium

$$M_A + M_B = P\Delta \tag{2.77}$$

Substituting Equation 2.76 into Equation 2.77 and simplifying

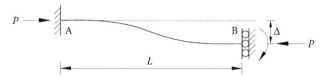

FIGURE 2.17 Beam-column with one fixed end and one guided end.

$$\frac{PL^2}{2EI} = s_{ii} + s_{ij}$$

or

$$\frac{k^2L^2}{2} = s_{ii} + s_{ij} \tag{2.78}$$

Upon substitution of the stability functions into Equation 2.78 and simplification, the following is obtained:

$$kL \sin kL = 0 \tag{2.79}$$

The lowest value of kL that satisfies Equation 2.79 is

$$kL = \pi$$

Then,

$$k^2L^2 = \pi^2 = \frac{PL^2}{EI} \tag{2.80}$$

that is, the critical load of the beam-column is

$$P_{cr} = \frac{\pi^2 EI}{L^2} \tag{2.81}$$

The critical load of this member can be obtained by solving the differential equation of equilibrium. The equilibrium equation of one segment of this member is

$$M_{int} - \frac{P\Delta}{2} + Py = 0 \tag{2.82}$$

Since

$$M_{int} = EIy'' \tag{2.83}$$

thus, the equilibrium equation can be written as

$$y'' + k^2 y = \frac{P\Delta}{2EI} \tag{2.84}$$

which has a general solution of the form

$$y(x) = C_1 \sin kx + C_2 \cos kx + \frac{\Delta}{2} \tag{2.85}$$

After imposing the boundary conditions, Equation 2.85 would lead to the same characteristic equation of the member, Equation 2.79.

Example 2.3

This example, Figure 2.18, describes a continuous beam-column with two outer ends fixed and an interior joint that prevents lateral displacement. Both column spans have the same flexural rigidity, EI, but are of different lengths, $0.5L$ and L. The rotations at ends A and C are zero; therefore, the slope-deflection equations of the moments at B are

$$M_{BA} = \frac{EI}{0.5L}\left(s_{ii}^{BA}\theta_B\right) = \frac{2EI}{L}\left(s_{ii}^{BA}\theta_B\right) \tag{2.86a}$$

$$M_{BC} = \frac{EI}{L}\left(s_{ii}^{BC}\theta_B\right) \tag{2.86b}$$

From equilibrium at joint B

$$M_{BA} + M_{BC} = 0 \tag{2.87}$$

Equations 2.86a and b and 2.87 yield the following relation:

$$2s_{ii}^{BA} + s_{ii}^{BC} = 0 \tag{2.88}$$

From Table 2.1

$$s_{ii}^{BA} = \frac{0.5kL\sin 0.5kL - (0.5kL)^2\cos 0.5kL}{2 - 2\cos 0.5kL - 0.5kL\sin 0.5kL} \tag{2.89a}$$

$$s_{ii}^{BC} = \frac{kL\sin kL - (kL)^2\cos kL}{2 - 2\cos kL - kL\sin kL} \tag{2.89b}$$

Upon substitution of s_{ii}^{BA} and s_{ii}^{BC} into Equation 2.88 and simplification the characteristic equation can be obtained, which upon solution yields

$$kL = 5.412$$

This gives the critical load of this continuous beam-column

$$P_{cr} = \frac{2.97\pi^2 EI}{L^2} \tag{2.90}$$

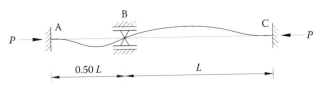

FIGURE 2.18   Continuous beam-column example.

2.4.2.1 Euler Load

The critical load of any beam-column of different boundary conditions can be determined by following the previous procedure using the slope-deflection equations. Alternatively, a solution can be obtained by forming the differential equation of equilibrium and then solving the equation imposing the appropriate boundary conditions. One of the interesting cases is a column of two hinged ends, Figure 2.19, for which the critical load is

$$P_{cr} = \frac{\pi^2 EI}{L^2} \tag{2.91}$$

The critical load of a column with two hinged ends is called Euler load, P_E, that is,

$$P_E = \frac{\pi^2 EI}{L^2} \tag{2.92}$$

2.4.3 THE K FACTOR AND THE ALIGNMENT CHARTS

2.4.3.1 K Factor

The critical load of any beam-column of any boundary condition, P_{cr}, can be written in a unified form as follows:

$$P_{cr} = \frac{\pi^2 EI}{L_b^2} \tag{2.93}$$

where L_b is the buckling length of the column, which is the distance between the inflection points of the beam-column in its deformed shape as a pin-pin column. From Equations 2.92 and 2.93, the column critical load, P_{cr}, can be related to Euler load, P_E, as follows:

$$P_{cr} = \frac{P_E}{(L_b/L)} = \frac{P_E}{K^2} \tag{2.94}$$

where the factor K is the ratio between the column buckling length, L_b, and its length, L. The value of K of the beam-column in Figure 2.16a is 0.7 ($P_{cr} = 2.041 P_E = P_E/(0.7)^2$ or $L_b = 0.7L$), while K is equal to 1.0 for the beam-column in Figure 2.17 and of course

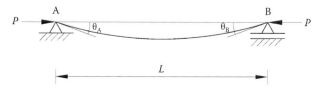

FIGURE 2.19 Pin-ended column.

TABLE 2.2
Theoretical K Values of Idealized Columns

(a)	(b)	(c)	(d)	(e)	(f)
0.5	0.7	1.0	1.0	2.0	2.0

Note: Buckled shape of column is shown by dashed line

---

for the pin-ended column. Different values of K, Table 2.2, are given for beam-columns of different boundary conditions, and the values range from 0.5 for a braced fixed ended column to 2.0 for a column of one fixed end and one free end. The value of K of the continuous beam-column in Figure 2.18 is 0.58 ($P_{cr} = 2.97P_E = P_E/(0.58)^2$ or $L_b = 0.58L$), which is approximately the average value of cases (a) and (b) in Table 2.2.

In the design of a structural system, where columns are not isolated members, the assessment of the critical load should follow a different approach. For instance, the critical load of a system can be determined by the slope-deflection equations or a second-order analysis. On the other hand, the value of the K factor can be assessed with the aid of the alignment charts that are derived from the slope-deflection method.

2.4.3.2 Alignment Charts

The alignment charts have been developed by Julian and Lawrence (1959) for a simplified assessment of the K factor. In the development of the charts the two models in Figure 2.20 are used for a system that is braced against side sway and an unbraced system. In both models, the following assumptions are adopted:

1. All columns are prismatic and behave elastically.
2. The axial forces in the beams are negligible.
3. All columns in a story buckle simultaneously.

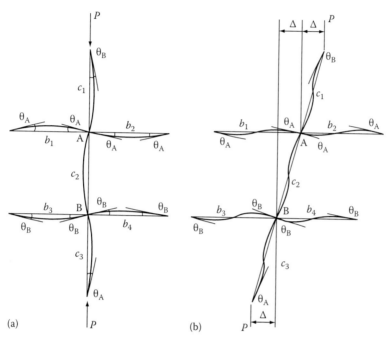

(a)

(b)

FIGURE 2.20 Subassemblage models: (a) subassemblage model for braced frame and (b) subassemblage model for unbraced frame.

4. At a joint, the restraining moment provided by the beams is distributed among the columns in proportion to their stiffnesses.
5. At buckling the girders are bent as shown in the models.

In both models in Figure 2.20, column c_2 (member AB) is the column in question. After writing the slope-deflection equations of either model and applying equilibrium at joints A and B, a relation of K is obtained in terms of two end stiffness parameters of the column, G_A and G_B, where

$$G_A = \frac{\sum_A (I/L)_c}{\sum_A (I/L)_b} = \frac{\sum \text{ of column stiffnesses meeting at joint A}}{\sum \text{ of beam stiffnesses meeting at joint A}} \qquad (2.95a)$$

$$G_B = \frac{\sum_B (I/L)_c}{\sum_B (I/L)_b} = \frac{\sum \text{ of column stiffnesses meeting at joint B}}{\sum \text{ of beam stiffnesses meeting at joint B}} \qquad (2.95b)$$

The indices b and c stand for beam and column, respectively, and the indices A and B stand for joints A and B, respectively. The obtained relation of K is expressed in a monograph form (alignment chart) for both cases of bracing conditions (Figure 2.21).

2.4.4 STABILITY ANALYSIS OF FRAMED STRUCTURES

The application of the slope-deflection method for stability analysis of a framed system is illustrated by the following two examples (Figure 2.22). The first example is a frame where sway is not allowed and the second example is a frame where sway is allowed.

Example 2.4

In this example it is required to determine the collapse load of the portal frame shown in Figure 2.22a, where sway is assumed to be prevented. The deflected shape of the frame is assumed as illustrated by the dashed lines. Upon writing the slope-deflection equations for the column joints A and B and eliminating θ_A since $M_A = 0$, the following equation is obtained:

$$M_{BA} = \frac{EI_c}{L_c}\left(s_{iic} - \frac{s_{ijc}^2}{s_{iic}}\right)\theta_B \tag{2.96}$$

The index c stands for column. Writing the slope-deflection equations for the beam joints B and C and eliminating θ_C since $\theta_C = -\theta_B$, the following equation is obtained:

$$M_{BC} = \frac{EI_b}{L_b}(s_{iib} - s_{ijb})\theta_B \tag{2.97}$$

The index b stands for beam. Assuming that the axial force in the beam is so small that it can be neglected, $s_{iib} = 4$ and $s_{ijb} = 2$; that is,

$$M_{BC} = \frac{2EI_b}{L_b}\theta_B \tag{2.98}$$

Upon applying equilibrium at joint B,

$$M_{BA} + M_{BC} = \frac{EI_c}{L_c}\left(s_{iic} - \frac{s_{ijc}^2}{s_{iic}}\right)\theta_B + \frac{2EI_b}{L_b}\theta_B = 0$$

leading to

$$\frac{EI_c}{L_c}\left(s_{iic} - \frac{s_{ijc}^2}{s_{iic}}\right) + \frac{2EI_b}{L_b} = 0 \tag{2.99}$$

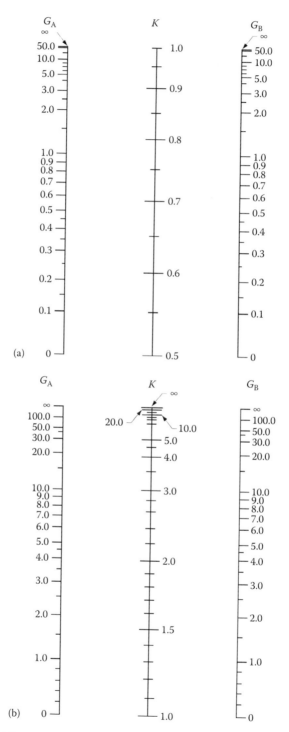

FIGURE 2.21 Alignment charts: (a) braced system and (b) unbraced system.

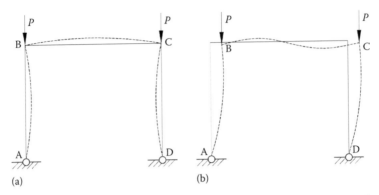

FIGURE 2.22 Examples of framed systems: (a) no side-sway and (b) side-sway allowed.

This equation is the characteristic equation of the frame. For the special case $I_c = I_b$ and $L_c = L_b$, Equation 2.99 becomes

$$S_{iic} - \frac{S_{ijc}^2}{S_{iic}} + 2 = 0 \tag{2.100}$$

By trial and error the value of kL that satisfies Equation 2.100 is

$$kL = \sqrt{\frac{P}{EI}} L = 3.59$$

which gives a value of the critical load

$$P_{cr} = 12.9 \frac{EI}{L^2} \tag{2.101}$$

Example 2.5

In this example it is required to determine the collapse load of the portal frame shown in Figure 2.22b, where sway is allowed. The deflected shape of the frame is assumed as illustrated by the dashed lines. Upon writing the slope-deflection equations for the column joints A and B and eliminating θ_A since $M_A = 0$, the following equation is obtained:

$$M_{BA} = \frac{EI_c}{L_c} \left[\left(S_{iic} - \frac{S_{ijc}^2}{S_{iic}} \right) \theta_B - \left(S_{iic} - \frac{S_{ijc}^2}{S_{iic}} \right) \frac{\Delta}{L_c} \right] \tag{2.102}$$

Writing the slope-deflection equations for the beam joints B and C and after eliminating θ_C, since the beam is in double curvature $\theta_C = \theta_B$, the following equation is obtained:

$$M_{BC} = \frac{EI_b}{L_b}(s_{iib} + s_{ijb})\theta_B \qquad (2.103)$$

Assuming that the axial force in the beam is so small that it can be neglected, $s_{iib} = 4$ and $s_{ijb} = 2$; that is,

$$M_{BC} = \frac{6EI_b}{L_b}\theta_B \qquad (2.104)$$

Upon applying equilibrium at joint B,

$$M_{BA} + M_{BC} = \frac{EI_c}{L_c}\left[\left(s_{iic} - \frac{s_{ijc}^2}{s_{iic}}\right)\theta_B - \left(s_{iic} - \frac{s_{ijc}^2}{s_{iic}}\right)\frac{\Delta}{L_c}\right] + \frac{6EI_b}{L_b}\theta_B = 0$$

leading to

$$\left(s_{iic} - \frac{s_{ijc}^2}{s_{iic}} + 6\frac{I_bL_c}{I_cL_b}\theta_B\right)\theta_B - \left(s_{iic} - \frac{s_{ijc}^2}{s_{iic}}\right)\frac{\Delta}{L_c} = 0 \qquad (2.105)$$

From the story shear equilibrium of the frame

$$\frac{M_{AB} + M_{BA} + P\Delta}{L_c} + \frac{M_{CD} + M_{DC} + P\Delta}{L_c} = 0 \qquad (2.106)$$

It should be noted that

$$M_{AB} = M_{DC} = 0 \quad \text{(hinged)} \qquad (2.107a)$$

$$M_{CD} = M_{BA} \quad \text{(antisymmetry)} \qquad (2.107b)$$

Upon substitution from the last equation in the shear equation in terms of M_{BA} and substitution of M_{BA} from Equation 2.102 and simplification, the following can be obtained:

$$\left(s_{iic} - \frac{s_{ijc}^2}{s_{iic}}\right)\theta_B - \left(s_{iic} - \frac{s_{ijc}^2}{s_{iic}} - \frac{P}{EI_c}L_c^2\right)\frac{\Delta}{L_c} = 0 \qquad (2.108)$$

Equations 2.105 and 2.108 are the frame equilibrium equations and can be written in matrix form as follows:

$$\begin{bmatrix} S+6\dfrac{I_b L_c}{I_c L_b} & -S \\[2ex] -S & S-\dfrac{PL_c^2}{EI_c} \end{bmatrix} \begin{pmatrix} \theta_B \\[1ex] \dfrac{\Delta}{L_c} \end{pmatrix} = \begin{pmatrix} 0 \\ 0 \end{pmatrix} \tag{2.109}$$

where

$$S = S_{iic} - \dfrac{S_{ijc}^2}{S_{iic}} \tag{2.110}$$

For the special case $I_c = I_b = I$ and $L_c = L_b = L$, Equation 2.109 becomes

$$\begin{bmatrix} S+6 & -S \\ -S & S-k^2L^2 \end{bmatrix} \begin{pmatrix} \theta_B \\[1ex] \dfrac{\Delta}{L} \end{pmatrix} = \begin{pmatrix} 0 \\ 0 \end{pmatrix} \tag{2.111}$$

where $k^2 = P/EI$. At bifurcation, both θ_B and Δ increase without bound and therefore in order for Equation 2.111 to be valid

$$\begin{vmatrix} S+6 & -S \\ -S & S-k^2L^2 \end{vmatrix} = 0 \tag{2.112}$$

Equation 2.112 is the characteristic equation of the frame. By trial and error the value of kL that satisfies this equation is

$$kL = \sqrt{\dfrac{P}{EI}}L = 1.35$$

which gives a value of the critical load

$$P_{cr} = 1.82\dfrac{EI}{L^2} \tag{2.113}$$

In the preceding examples, the slope-deflection method has been used to determine the critical load of a framed system. In case this calculation regime is difficult to perform as in the case of large structures, codes of practice adopt approximate methods in order to account for the second-order stability effects. These methods are commonly known as the amplification (or magnification) methods, which are presented next.

2.4.5 Amplification Factors for Second-Order Effects

With reference to Figure 2.23, structural stability has two fundamental effects: the first is the $P-\Delta$ effect and the second is the $P-\delta$ effect. The first effect may have a softening or stiffening effect on structural members; nevertheless, such an effect is

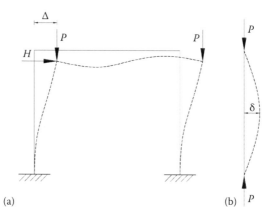

FIGURE 2.23   (a) P–Δ effect and (b) P–δ effect.

generally detrimental for columns. With reference to Figure 2.23a, the P–Δ effect is associated with additional loads on columns due to the interaction between the load and nodal displacements (the shift of the load from its original position because of structural deformations). The lateral force, H, acting on the frame alone causes a lateral drift, Δ; however, with the presence of vertical forces, ΣP, this drift and the overturning moment will increase further as a result of the interaction between the vertical forces and the lateral drift.

The member stability (P–δ effect), Figure 2.23b, will have a softening effect on a slender member if P is compression, and a stiffening effect if P is tension. The softening effect arises from the additional moments due to the interaction between the axial compressive force and the lateral deflection of the member, δ. If the axial force is tension, the interaction between this force and the member lateral deflection will reduce the moments along the column span. Usually columns are subjected to compression rather than tension and therefore the P–δ effect is critical.

In order to approximately assess the stability effects without having to carry out a slope-deflection analysis or a second-order analysis, the moment amplification (or magnification) has been developed. In this method, the P–Δ effect and the P–δ effect are estimated independently as illustrated in the following sections.

2.4.5.1 P–Δ Effect

When the lateral forces ΣH act on a frame and cause a primary lateral deflection, Δ_i, which can be obtained from a first-order analysis, the vertical forces will inter-act with this deflection. As a result, additional lateral deflection and additional moments take place, which is called the P–Δ effect. This effect can be estimated by different methods such as the story magnifier method and the multiple-column magnifier method (Chen and Lui, 1991). In the former method, it is assumed that each story behaves independently of other stories and that the additional moment in the columns caused by the P–Δ effect is equivalent to that caused by a lateral force of $\Sigma P\Delta/h$ (h is the story height). Hence, the sway stiffness of the story can be determined as

$$S_F = \frac{\Sigma H}{\Delta_i} = \frac{\Sigma H + \Sigma P \Delta / h}{\Delta} \tag{2.114}$$

where

ΣH is the sum of all story horizontal forces producing Δ_i which is the first-order translational deflection of the story under consideration
ΣP is the axial load on all columns in that story

Solving Equation 2.114 for the total deflection Δ,

$$\Delta = \left(\frac{1}{1 - \Sigma P \Delta_i / \Sigma H h} \right) \Delta_i \tag{2.115}$$

Since every story is assumed to behave independently of other stories, the sway moment as a result of the story swaying is proportional to the lateral deflection of the story. Therefore, the total moment due to sway and second-order effect (in terms of the primary story moment, M_{isway}) is

$$M = \left(\frac{1}{1 - \Sigma P \Delta_i / \Sigma H h} \right) M_{isway} \tag{2.116}$$

2.4.5.2 P–δ Effect

In order to estimate the P–δ effect, the beam-column in Figure 2.24 is considered. The vertical loads and end moments produce primary moment, M_i, and primary deflection, v_i. As a result of the interaction between the axial force, P, and the primary deflection, additional moment, M_{ii}, and additional deflection, v_{ii}, which is called the P–δ effect, take place. The total moment, M, and the total deflection, v, are thus,

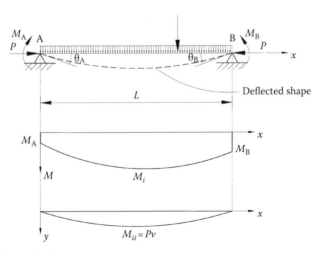

FIGURE 2.24 P–δ effect.

$$M = M_i + M_{ii} \tag{2.117a}$$

$$v = v_i + v_{ii} \tag{2.117b}$$

For an approximate assessment of this effect (Chen and Lui, 1991), assume that M_{ii} has the shape of half sine wave and that v_{max} $(= \delta = \delta_i + \delta_{ii})$ occurs at midspan:

$$M_{ii} = P\delta \sin \frac{\pi x}{L} \tag{2.118}$$

The moment M_{ii} is related to the second derivative of v_{ii} by

$$v_{ii}'' = -\frac{M_{ii}}{EI} \tag{2.119}$$

The negative sign in Equation 2.119 is because the moment increases while the slope decreases and vice versa. From Equations 2.118 and 2.119,

$$v_{ii}'' = -\frac{P\delta}{EI} \sin \frac{\pi x}{L} \tag{2.120}$$

Upon double integrating Equation 2.120 and introducing the boundary conditions ($v_{ii} = 0$ at $x = 0, L$) and calculating $\delta_{ii} = v_{ii}$ at midspan,

$$\delta_{ii} = -\frac{P\delta}{P_E} \tag{2.121}$$

Since

$$\delta = \delta_i + \delta_{ii} \tag{2.122}$$

substituting Equation 2.121 into Equation 2.122 and solving for δ,

$$\delta = \frac{1}{1 - P/P_E} \delta_i \tag{2.123}$$

If the maximum primary moment is assumed to take place in the midspan, the following equation for the maximum moment can be obtained:

$$M_{max} = \frac{1 + \psi P/P_E}{1 - P/P_E} M_{i\,max} \tag{2.124}$$

where

$$\psi = \frac{\delta_i P_E}{M_{i\max}} \tag{2.125}$$

Equation 2.124 can be written in the following form:

$$M_{\max} = \frac{C_m}{1 - P/P_E} M_{i\max} \tag{2.126}$$

where

$$C_m = 1 + \frac{\psi P}{P_E} \tag{2.127}$$

It should be mentioned that the value of C_m in Equation 2.127 is applicable only when the maximum primary moment, $M_{i\max}$, takes place at or near midspan.

Example 2.6

For demonstration, the amplification factor methods are applied to the frame shown in Figure 2.25. The axial rigidity and flexural rigidity of all frame members have the same values, $EA = 2.619 \times 10^6$ kN and $EI = 3.852 \times 10^4$ kN m^4, respectively. From first-order analysis, the lateral displacements of joints B and C are $\Delta_B = 26.41$ mm and $\Delta_C = 26.22$ mm, respectively, and the maximum deflection of member AB is $\delta_{iAB} = 7.33$ mm and of member DC is $\delta_{iDC} = 10.373$ mm.

For the P–Δ effect, $\Sigma P\Delta_i/\Sigma Hh = 0.05263$. From Equation 2.116, $M = 1.056 M_{isway}$; that is, the end moment of column AB at B is $M = 1.056 \times 115.3 = 121.76$ kN m, and the end moment of column DC at C is $M = 1.056 \times 249.8 = 263.8$ kN m.

For the P–δ effect, $P_E = 152.2 \times 10^2$ kN for either column AB or column DC. For column AB, $P/P_E = 0.0123$ and $M_{i\max} \approx 90.4$ kN m. From Equation 2.125, $\psi_{AB} = 1.234$; thus, from Equation 2.124, $M_{\max} = 1.028 M_{i\max}$. For column DC, $P/P_E = 0.014$ and $M_{i\max} \approx 125.0$ kN m, $\psi_{DC} = 1.263$ and $M_{\max} = 1.0282 M_{i\max}$.

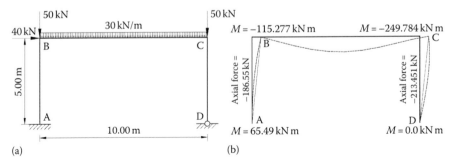

FIGURE 2.25 Frame Example 2.6: (a) dimensions and loads and (b) deformed shape and straining actions.

2.5 THEORY OF PLATES

The generalized stress–generalized strain relations of the thin plate element under bending have been derived from the three sets of conditions: equilibrium, compatibility, and constitutive law. By this, a major simplification has been introduced to the solution of the bending problem: dealing with 2 variables instead of 15 and treating a whole element instead of a field point. Expanding the discussion of this topic by considering the equilibrium of the element shown in Figure 2.26, the following differential equation of equilibrium can be derived:

$$\frac{\partial^4 w}{\partial x^4} + 2\frac{\partial^4 w}{\partial x^2 \partial y^2} + \frac{\partial^4 w}{\partial y^4} = \frac{p}{D} \tag{2.128}$$

where p is the load intensity.

In this section, the concept of generalized stress and generalized strain with the three sets of conditions is employed to derive different solutions of thin plate bending problems.

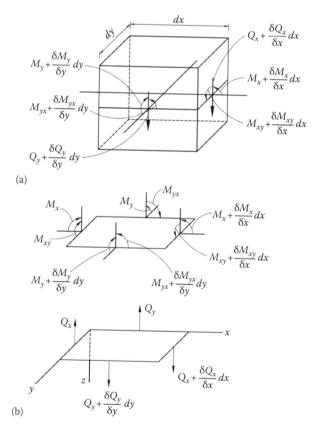

(a)

(b)

FIGURE 2.26 (a) Plate element equilibrium and (b) stress resultants.

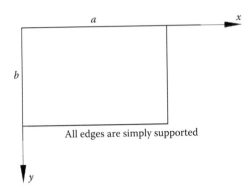

FIGURE 2.27 Simply supported plate.

The simply supported plate shown in Figure 2.27 is subjected to a sinusoidal load given by

$$p(x,y) = p_o \sin\frac{\pi x}{a} \sin\frac{\pi y}{b} \qquad (2.129)$$

where a and b are the plate dimensions in the x and y directions, respectively. Upon substitution of the load formula in the differential equation of equilibrium,

$$\frac{\partial^4 w}{\partial x^4} + 2\frac{\partial^4 w}{\partial x^2 \partial y^2} + \frac{\partial^4 w}{\partial y^4} = \frac{p_o}{D}\sin\frac{\pi x}{a}\sin\frac{\pi y}{b} \qquad (2.130)$$

From kinematics,

$$w = 0 \text{ and } \frac{\partial^2 w}{\partial x^2} = 0 \quad \text{at } x = 0 \text{ and } x = a \qquad (2.131a)$$

$$w = 0 \text{ and } \frac{\partial^2 w}{\partial y^2} = 0 \quad \text{at } y = 0 \text{ and } y = b \qquad (2.131b)$$

In order to solve this plate problem, the deflection expression must satisfy the aforementioned kinematic relations and should have the same form as the load expression. Hence, the deflection should have the following form:

$$w = C\sin\frac{\pi x}{a}\sin\frac{\pi y}{b} \qquad (2.132)$$

where C is a constant that can be obtained upon substitution of Equation 2.132 into Equation 2.130:

$$C = \frac{p_o}{D\pi^4\left(\left(1/a^2\right)+\left(1/b^2\right)\right)^2} \qquad (2.133)$$

or

$$w = \frac{p_o}{D\pi^4\left(\left(1/a^2\right)+\left(1/b^2\right)\right)^2} \sin\frac{\pi x}{a}\sin\frac{\pi y}{b} \qquad (2.134)$$

The stress resultants, Q_x, Q_y, M_x, M_y, M_{xy}, and M_{yx} can be obtained from

$$Q_x = -D\frac{\partial}{\partial x}\left(\frac{\partial^2 w}{\partial x^2}+\frac{\partial^2 w}{\partial y^2}\right)$$

$$Q_y = -D\frac{\partial}{\partial y}\left(\frac{\partial^2 w}{\partial x^2}+\frac{\partial^2 w}{\partial y^2}\right)$$

$$M_x = -D\left(\frac{\partial^2 w}{\partial x^2}+v\frac{\partial^2 w}{\partial y^2}\right) \qquad (2.135)$$

$$M_y = -D\left(v\frac{\partial^2 w}{\partial x^2}+\frac{\partial^2 w}{\partial y^2}\right)$$

$$M_{xy} = (1-v)D\left(\frac{\partial^2 w}{\partial x\partial y}\right) = -M_{yx}$$

Of course, the strains and stresses at any point can be obtained if needed:

$$\varepsilon_x = -z\frac{\partial^2 w}{\partial x^2}$$

$$\varepsilon_y = -z\frac{\partial^2 w}{\partial y^2}$$

$$\gamma_{xy} = -2z\frac{\partial^2 w}{\partial x\partial y}$$

$$\sigma_x = \frac{E}{1-v^2}(\varepsilon_x+v\varepsilon_y) \qquad (2.136)$$

$$\sigma_y = \frac{E}{1-v^2}(v\varepsilon_x+\varepsilon_y)$$

$$\tau_{xy} = \frac{E\gamma_{xy}}{2(1+v)}$$

Let us examine the same simply supported plate in Figure 2.27 if subjected to a general distributed load $p(x,y)$. The load can be represented by double Fourier series (Navier) as follows:

$$p(x,y) = \sum_{m=1}^{\infty} \sum_{n=1}^{\infty} p_{mn} \sin \frac{m\pi x}{a} \sin \frac{n\pi y}{b} \tag{2.137}$$

In order to obtain p_{mn}, both sides of Equation 2.137 are multiplied by $\sin(m'\pi x/a)$ $\sin(n'\pi y/b)$; then, both sides of the equation are integrated, which leads to

$$p_{mn} = \frac{4}{ab} \int_0^a \int_0^b p(x,y) \sin \frac{m\pi x}{a} \sin \frac{n\pi y}{b} \, dx dy \tag{2.138}$$

From the previous solution of the plate under sinusoidal load and upon applying the principle of superposition, a solution of a simply supported plate under general loading condition can be obtained:

$$w = \frac{1}{D\pi^4} \sum_{m=1}^{\infty} \sum_{n=1}^{\infty} \frac{p_{mn}}{\left((m/a)^2 + (n/b)^2\right)^2} \sin \frac{m\pi x}{a} \sin \frac{n\pi y}{b} \tag{2.139}$$

If $p(x,y) = p_o$, then

$$p_{mn} = \frac{4p_o}{ab} \int_0^a \int_0^b \sin \frac{m\pi x}{a} \sin \frac{n\pi y}{b} = \frac{16p_o}{\pi^2 mn} \tag{2.140}$$

and

$$w = \frac{16p_o}{\pi^6 D} \sum_{m=1}^{\infty} \sum_{n=1}^{\infty} \frac{\sin(m\pi x/a) \sin(n\pi y/b)}{mn\left((m/a)^2 + (n/b)^2\right)^2} \tag{2.141}$$

$m, n = 1,3,5,\ldots$ since the terms of equations vanish for even values of m or n.

2.6 THEORY OF SHELLS

The generalized stress–generalized strain relations of thin cylindrical shells under axisymmetric loading have been derived from the three sets of conditions: equilibrium, compatibility, and constitutive law. Thus, the number of variables is reduced from 15 to 3, which is a remarkable simplification. In addition, a solution of a whole structural element or structure can be obtained at once. The discussion is extended in this section in order to show how the concept of generalized stress–generalized strain along with the three basic sets of conditions are employed to obtain solutions of different problems of thin cylindrical shells under axisymmetric loading. The discussion starts with the derivation of the general solution of this class of problems.

The following equilibrium equations have been derived earlier in Section 2.2.4:

$$\frac{dQ_x}{dx} + \frac{N_\theta}{r} = p \tag{2.142}$$

$$\frac{dM_x}{dx} - Q_x = 0 \tag{2.143}$$

Upon differentiating Equation 2.143 with respect to x, and substituting for dQ_x/dx from (2.142), the following equation can be obtained:

$$\frac{d^2M_x}{dx^2} + \frac{N_\theta}{r} = p \tag{2.144}$$

Upon substitution from Equation 2.46 for N_θ, from Equation 2.47 for M_x and $\varphi_x = \partial^2 w/\partial x^2$ into Equation 2.144, the following differential equation of equilibrium can be obtained:

$$\frac{\partial^4 w}{\partial x^4} + \frac{Etw}{Dr^2} = \frac{p}{D} \tag{2.145}$$

Define

$$4\beta^4 = \frac{Et}{Dr^2} \tag{2.146}$$

Then, Equation 2.145 can be written as follows:

$$\frac{\partial^4 w}{\partial x^4} + 4\beta^4 w = \frac{p(x)}{D} \tag{2.147}$$

Equation 2.147 is the differential equation of equilibrium of thin cylindrical shells subjected to axisymmetric loading. The general solution of this equation has the form

$$w = e^{\beta x}(C_1 \cos \beta x + C_2 \sin \beta x) + e^{-\beta x}(C_3 \cos \beta x + C_4 \sin \beta x) + f(x) \tag{2.148}$$

The term $f(x)$ is the particular solution and the constants C_1 to C_4 can be obtained from the boundary conditions.

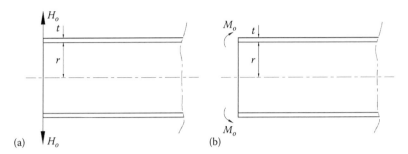

FIGURE 2.28 Example of a long circular cylindrical shell: (a) shear at the edge and (b) moment at the edge.

In the following, the concept of generalized stress–generalized strain along with the three basic sets of conditions is used to obtain solutions for long thin cylindrical shells under different loading conditions. In the following example, a long cylinder is solved for a shear force at the edge, Figure 2.28a, and for a bending moment at the edge, Figure 2.28b. For both problems, the function $f(x)$ is set equal to zero since there is no applied pressure. In addition, the deflection associated with the term $e^{\beta x}$ tends to approach infinity as x tends to a large value. This contradicts the real behavior of the cylinder under the shown edge load, in which the deflection and moments dissipate as x increases. Therefore, the constants C_1 and C_2 of Equation 2.148 should vanish; thus, the equation reduces to

$$w = e^{-\beta x}(C_3 \cos \beta x + C_4 \sin \beta x) \tag{2.149}$$

In order to obtain the constants C_3 and C_4, the boundary conditions for each problem are introduced. For the problem in Figure 2.28a, the conditions are

$$M_x(x = 0) = 0 \tag{2.150a}$$

$$Q_x(x = 0) = H_o = \frac{dM_x}{dx}\bigg|_{x=0} \tag{2.150b}$$

For the problem in Figure 2.28b, the conditions are

$$M_x(x = 0) = M_o \tag{2.151a}$$

$$Q_x(x = 0) = 0 = \frac{dM_x}{dx}\bigg|_{x=0} \tag{2.151b}$$

The solution of the problem in Figure 2.28a is

$$M_x = \frac{H_o}{\beta} e^{-\beta x} \sin \beta x$$

$$Q_x = \sqrt{2} H_o e^{-\beta x} \cos\left(\beta x + \frac{\pi}{4}\right)$$

$$N_\theta = 2 H_o \beta r e^{-\beta x} \cos \beta x$$

$$M_\theta = \nu M_x \qquad\qquad (2.152)$$

$$w = \frac{H_o}{2 D \beta^3} e^{-\beta x} \cos \beta x$$

$$\frac{dw}{dx} = \frac{-H_o}{\sqrt{2} D \beta^2} e^{-\beta x} \sin\left(\beta x + \frac{\pi}{4}\right)$$

The solution of the problem in Figure 2.28b is

$$M_x = \sqrt{2} M_o e^{-\beta x} \cos\left(\beta x - \frac{\pi}{4}\right)$$

$$Q_x = -2 M_o e^{-\beta x} \sin \beta x$$

$$N_\theta = -2\sqrt{2} M_o \beta^2 r e^{-\beta x} \sin\left(\beta x - \frac{\pi}{4}\right)$$

$$M_\theta = \nu M_x \qquad\qquad (2.153)$$

$$w = \frac{-M_o}{\sqrt{2} D \beta^2} e^{-\beta x} \sin\left(\beta x - \frac{\pi}{4}\right)$$

$$\frac{dw}{dx} = \frac{M_o}{D \beta} e^{-\beta x} \sin\left(\beta x - \frac{\pi}{2}\right)$$

2.7 FINITE ELEMENT

The FE method is based on three concepts: (1) the concept of generalized stresses and generalized strains; (2) equilibrium is justified through the principle of virtual displacement; and (3) compatibility is achieved through the assumed shape function. As a consequence, it has become possible to analyze any structure of any geometry and of different materials.

In this section, the solution of a few examples are presented in order to show (1) the level of accuracy that can be reached using the FE method and (2) how structural behavior can be reflected in the FE solution and thus to verify some mechanics

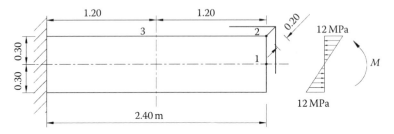

FIGURE 2.29 Example of a cantilever beam.

assumptions. In all examples, the elasticity modulus, $E=25,000\,\text{MPa}$, and Poisson's ratio, $\nu=0.2$, are used.

The first example is a cantilever beam subjected to an end moment, Figure 2.29, where the exact solution can be obtained from mechanics (Section 2.3.1). The rotation, $\theta(x)$, is

$$\theta(x) = \frac{M}{EI}(x) \tag{2.154}$$

and the deflection, $v(x)$, is

$$v(x) = \frac{M}{EI}\left(\frac{x^2}{2}\right) \tag{2.155}$$

The moment $M=1.44\times 10^8\,\text{N mm}$ and the flexural rigidity $EI=9\times 10^{12}\,\text{N mm}^2$; thus, $\theta(x=L)=0.0384\,\text{rad}$ and $v(x=L)=46.08\,\text{mm}$. Hence, the vertical deflection at 1 and 2 is, $v_1=v_2=46.08\,\text{mm}$ upward, and the horizontal displacement at 1 is $u_1=0$, while the horizontal displacement at 2 is calculated from the rotation as $u_2=-11.52\,\text{mm}$. From the FE solution of this problem it is found that, $u_1=0$, $u_2=-11.51\,\text{mm}$, $v_1=46.12\,\text{mm}$, and $v_2=45.98\,\text{mm}$. When these FE results are compared with the exact values, it is evident how much accurate the FE results are when using proper meshing and modeling. Also, from the FE results the strain distribution at section 3 is almost linear, which conforms to the simplification of Bernoulli's hypothesis of beams.

The second example, Figure 2.30, is a beam with different values of shear span to depth ratio, a/d. In this example, the division line between a deep beam and a short beam can be noticed from the strain distribution. From the FE results in Table 2.3, for a shear span to depth ratio $a/d \leq 2.0$, the beam behaves as a deep beam, which agrees with St. Venant's principle. Also, the stress trajectories show the arch action in deep beams and the truss model in ordinary beam.

The third example is a simply supported rectangular plate, Figure 2.27, with dimensions $a=4.00\,\text{m}$, $b=3.00\,\text{m}$, and $t=100\,\text{mm}$. The plate is subjected to a uniform load $p=10.0\,\text{kN/m}^2$. The exact value of maximum deflection obtained from mechanics (Section 2.5) was $w=2.54\,\text{mm}$ and from FE was $2.50\,\text{mm}$.

It should be mentioned that FE is advantageous not only in analyzing any problem of any shape but also in handling discontinuity problems, for example, openings,

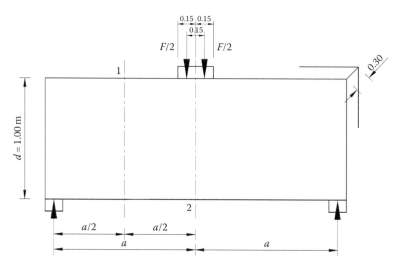

FIGURE 2.30 Example of a beam of different shear span to depth ratios.

bends and stress concentration, and irregular boundary conditions. The fourth example shown in Figure 2.31 is a deep beam with a central opening. From the FE results illustrated in the figure, the stress distribution at the central section (section 1) is nonlinear and the stress trajectories agree with the logical flow of forces. The same observations are valid for the deep beam with an eccentric opening (Figure 2.32). Another example shown in Figure 2.33 is a cantilever with a dapped end, where the illustrated FE results reflect the behavior of such a discontinuity region.

In conclusion, the FE can analyze any structure of any shape and of any boundary condition. If the FE can be considered "exact," which means the best possible solution, it can be used for calibration and verification, for instance, verification of kinematic assumptions such as Bernoulli's assumption. It can illustrate the limits of simplified models, for example, beam model, and the reliability of simplification, for example, Bernoulli's hypothesis. The FE can handle those problems where simplified models are not valid, for example, openings in beams, abrupt change of thickness, and discontinuity in general.

2.8   THE ALLOWABLE STRESS AS A BASIS FOR DESIGN

There is a fundamental two-stage process in structural design operation: first, the forces acting on each structural member must be defined, and second, the load-carrying capacity of each member must be determined. The first stage involves an analysis of the stresses acting within the structural members and the second involves knowledge of the load-carrying capacity of the structural members. In allowable stress design, the first stage is based on a linear elastic analysis while the second stage is based on full-scale tests of structural members reduced by an acceptable safety factor to the allowable stress level as specified by specifications.

This was the era of great advancement in the use of the mathematical theory of elasticity for structural design, but one that placed too much emphasis on the elastic

TABLE 2.3

FE Results of the Example Beam of Different Shear Span to Depth Ratios (Figure 2.30)

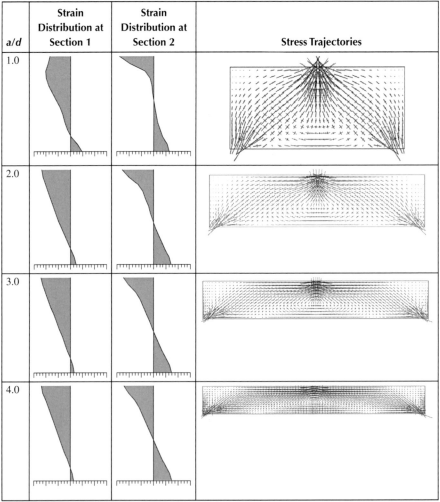

a/d	Strain Distribution at Section 1	Strain Distribution at Section 2	Stress Trajectories
1.0			
2.0			
3.0			
4.0			

response of an idealized version of the real structure to normal working loads. It turned the engineers so far toward the analysis of the working load range that the much more important task of insuring the safety of a design against collapse or failure was submerged till the emerging development of the theory of plasticity in the 1950s.

In order to arrive at a reasonable estimate of the safety factor, *SF*, the two diagrams in Figure 2.34 are examined; one diagram is the idealized stress–stress relation based on linear elasticity theory and the other is the actual behavior of the material. However, at the time of the design method initiation, the material curve

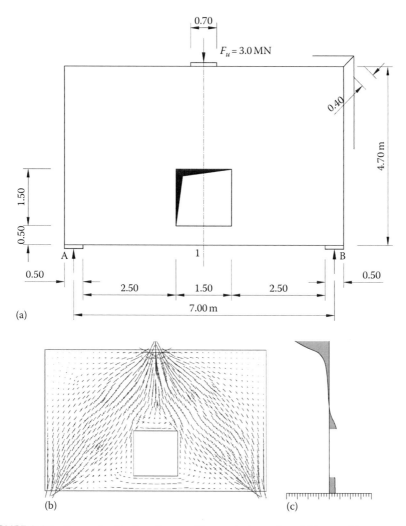

FIGURE 2.31 Example of a deep beam with a central opening: (a) beam; (b) stress trajectories; and (c) stress distribution at section 1.

could be measured up to the first yield point, while the strength beyond this point was not possible to be predicted and therefore was neglected. Hence, safety would have been related to this first yield point; that is, the allowable stress, σ_{all}, would have been a fraction of this yield stress, σ_y:

$$\sigma_{all} = \frac{\sigma_y}{SF} \tag{2.156}$$

The value of the safety factor would be based on the type of material (steel, concrete, etc.), type of straining action (flexure, axial, shear, etc.), type of load and load

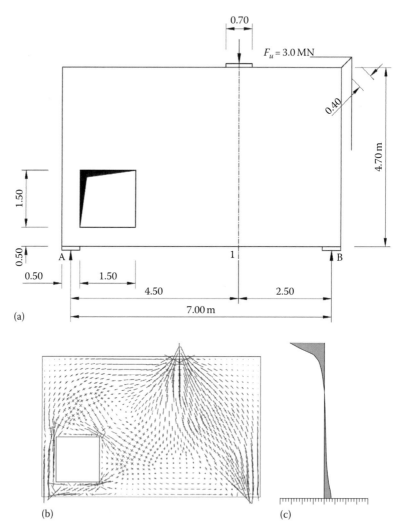

FIGURE 2.32 Example of a deep beam with an eccentric opening: (a) beam; (b) stress trajectories; and (c) stress distribution at section 1.

combination, type of structure, etc. The values of safety factor were defined through time with the accumulation of practical experience.

The allowable stress design method had been employed reasonably well in traditional structures as a result of the accumulative experience particularly under normal loading conditions. However, safety rules based on experience work well only for designs lying within the scope of that experience. They cannot be relied on outside of that range, for example, for the case of nontraditional structures, for example, offshore structures, poles, and nuclear facilities. In other words, structures susceptible to abnormal loads, for example, extreme wind, major earthquakes, and explosive loads, could not be handled properly with this method. Therefore, it was a logical

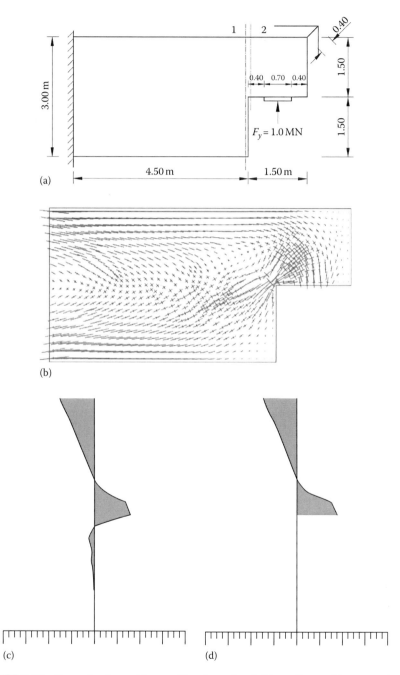

FIGURE 2.33 Example of a cantilever with a dapped end: (a) cantilever; (b) stress trajectories; (c) strain distribution at section 1; and (d) strain distribution at section 2.

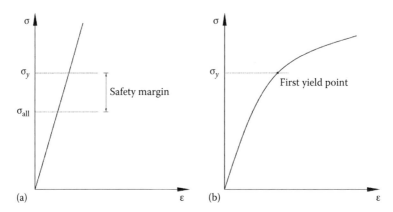

FIGURE 2.34 Idealized and actual material stress–strain relation: (a) idealized and (b) actual.

future development to establish more reliable and more rational inelastic second-order analysis methods for design that can avoid the drawbacks of the allowable stress design method. This further development of analysis and design methods is described in Chapter 3.

2.9 HISTORICAL SKETCH

Modern theory of elasticity began when the French mathematician Jacob Bernoulli combined equilibrium equations with Hooke's law (1678) to obtain the differential equation of the *elastica*, in 1705, that is, the curve assumed by the deformed axis of the beam. In 1821, 116 years after Bernoulli, the general equations of equilibrium and vibration of elastic solids were formulated by Navier, which led to the formulation of the linear theory of elasticity by A.L. Cauchy in 1822. This linear theory of elasticity remains virtually unchanged to the present day. During this period, following the discovery of Hooke's law, the growth of the science of elasticity proceeded from the synthesis of solutions of special problems. This began the development of flexural theory of beams, theory of torsion, theory of stability of columns, and some results on bending and vibration of plates in the early nineteenth century.

This chapter represents an attempt to present concisely several aspects of the theory of elasticity from a unified point of view and to indicate those familiar methods of solution of the field equations of elasticity that are familiar to structural engineers in particular. To this end, we must single out and stress the contributions to the engineering theory by the Russian elastician, S. P. Timoshenko, a great teacher, and whose work was widely known for its elegance and importance.

It was during his years in the United States that Timoshenko made the major part of his contributions and writings to the theory of elasticity and to its application to the design of engineering structures and components. Timoshenko, his colleagues, and his students became internationally known for their pioneering work in elasticity, especially those classical textbooks used widely in engineering practice and university teaching in particular. Timoshenko's unparalleled monographs on the applications of elasticity to engineering structures include

1. *Applied Elasticity*, D. Van Nostrand Company, Inc., New York, 1925 (with J. M. Lessells).
2. *Vibration Problems in Engineering*, D. Van Nostrand Company, Inc., New York, 1928, 1937 and 1955 (with D. H. Young).
3. *Strength of Materials, Part I, Elementary Theory and Problems*, D. Van Nostrand Company, Inc., Princeton, NJ, 1930, 1940, and 1955.
4. *Strength of Materials, Part II, Advanced Theory and Problems*, D. Van Nostrand Company, Inc., Princeton, NJ, 1930, 1941, and 1956.
5. *Theory of Elasticity*, McGraw-Hill Book Company, New York, 1934 and 1951 (with J. N. Goodier).
6. *Elements of Strength of Materials*, D. Van Nostrand Company, Inc., Princeton, NJ, 1935, 1940 and 1949 (with G. H. McCullough), 1962 (with D. H. Young).
7. *Theory of Elastic Stability*, McGraw-Hill Book Company, New York, 1936 and 1961 (with J. M. Gere).
8. *Engineering Mechanics*, McGraw-Hill Book Company, New York, 1937, 1940, 1951 and 1956 (with D. H. Young).
9. *Theory of Plates and Shells*, McGraw-Hill Book Company, New York, 1940, 1959 (with S. Woinowsky-Krieger).
10. *Theory of Structures*, McGraw-Hill Book Company, New York, 1945 and 1965 (with D. H. Young).
11. *Advanced Dynamics*, McGraw-Hill Book Company, New York, 1948 (with D. H. Young).
12. *History of Strength of Materials*, McGraw-Hill Book Company, New York, 1953.
13. *Engineering Education in Russia*, McGraw-Hill Book Company, New York, 1959.
14. *The Collected Papers of Stephen P. Timoshenko*, 1953, McGraw-Hill Book Company, New York.
15. *As I Remember*, D. Van Nostrand Company, Inc., Princeton, NJ, 1968.
16. *Stephen Timoshenko 60th Anniversary Volume, Contributions to the Mechanics of Solids Dedicated to Stephen Timoshenko by His Friends*, The MacMillan Company, New York, 1938.

S. P. Timoshenko was born on December 22, 1878 in Ukraine to parents who took great care of their children. He was self-motivated and since his childhood had a dream to become a structural engineer. After his graduation from Realgymnasium at the age of 18, Timoshenko obtained most of his early education and practical experience in Russia. In his early life he had two trips to western Europe, which had a great influence on his life and career. In his 1900 trip he visited Germany, Belgium, France, and Switzerland, where he realized how these countries were ahead of Russia in both culture and industrial development. During his 1904 trip he was very much impressed with August Föppl at Munich Polytechnical Institute.

In the academic year 1903–1904 he read English Love's *Theory of Elasticity* and Lord Rayleigh's *Theory of Sound*. These two books impressed him immensely and had a considerable influence on his scientific work. Timoshenko went twice to the University of Göttingen in Germany for a few months in 1905 and 1906, where he took some courses in mathematics and mechanics and investigated the problem of lateral buckling of I-beams. Timoshenko had his first teaching opportunity at Kiev Polytechnicum, where he commenced lecturing in January 1907. During his whole teaching career, and different from what was followed at his time, he used to start with the simplest problems and move gradually to the more complicated ones, which was much appreciated by his students.

In 1920, Timoshenko escaped to Zagreb because of the long suffering for many years under the Bolshevik regime and the World War I and because he was politically in danger. In Zagreb, he was satisfied with his scientific achievements; however, he decided to start his academic life in the United States in 1922. Nevertheless, he had only a chance to work in industry, first in a vibration company in Philadelphia for a few months and afterward in Westinghouse Company in Chicago. With many research accomplishments he stayed in Westinghouse for 5 years until he had the opportunity to return to academic work as he always hoped. In 1927, he joined the University of Michigan where he achieved teaching and research work of exceptional level such that his ideas and attitude toward mechanics became widespread throughout the United States. In 1936, he joined Stanford University until his formal retirement in 1944; however, he continued his academic activities at Stanford but with a reduced scale and with concentration on writing and revision of text books.

REFERENCES

Chen, W. F. and Han, D. J., 1988, *Plasticity for Structural Engineers*, Springer-Verlag, New York.

Chen, W. F. and Lui, E. M., 1987, *Structural Stability—Theory and Implementation*, Elsevier, New York.

Chen, W. F. and Lui, E. M., 1991, *Stability Design of Steel Frames*, CRC Press, Boca Raton, FL.

Jawad, M. H., 1994, *Theory and Design of Plate and Shell Structures*, Chapman & Hall, New York.

Kassimali, A., 2005, *Structural Analysis*, 3rd edn., Thomson, Toronto, Ontario, Canada.

Kelkar, V. S. and Sewell, R. T., 1987, *Fundamentals of the Analysis and Design of Shell Structures*, Prentice-Hall, Inc., Englewood Cliffs, NJ.

Lively, R. K., 1975, *Matrix Methods of Structural Analysis*, 2nd edn., Pergamon Press, New York.

The Collected Papers of Stephen P. Timoshenko, 1953, McGraw-Hill Book Company, New York.

Timoshenko, S. P., 1953, *History of Strength of Materials*, McGraw-Hill Book Company, New York.

Timoshenko, S. P., 1968, *As I Remember*, D. Van Nostrand Company, Inc., Princeton, NJ, Translated from Russian by R. Addis.

Timoshenko, S. P. and Goodier, J. N., 1970, *Theory of Elasticity*, 3rd edn., McGraw-Hill Book Company, New York.

Timoshenko, S. P. and Woinowsky-Krieger, S., 1959, *Theory of Plates and Shells*, McGraw-Hill Book Company, New York.

3 The Era of Plasticity

3.1 FUNDAMENTALS OF PLASTICITY

3.1.1 INTRODUCTION

Similar to the theory of elasticity, the theory of plasticity was founded on bold assumptions based on observations of experiments followed with idealizations and simplifications. The early experiments of Tresca (1870) in the 1860s established the concept that large plastic deformation is shear deformation governed primarily by shear stress. It took a great insight to ignore the time effects to start the mathematical theory of plasticity. The choice of maximum shear stress as the critical shear stress for the Tresca yield criterion for metallic material started the development of the theory of metal plasticity. Similarly, the choice of a normal-stress-dependent maximum shear stress as a failure criterion for granular media laid the foundation of Coulomb's (1773) failure criterion, proposed much earlier for soils, for the subsequent development of the lateral earth pressure theory in 1776 by the French physicist Charles-Augustin de Coulomb who began the modern theory of soil mechanics.

As a result, a yield or failure criterion combined with the equations of equilibrium has been used to solve some useful and interesting "statically determinate" types of structure problems. In soil mechanics, this type of solution is known as the "limit equilibrium" solution (Sokolovsky, 1946). However, the more recent development of the theory of elasticity has made it clear that plastic stress–strain relations are required for a complete continuum-mechanics solution in which structural engineering is a branch of its application. The plastic stress–strain relations proposed by St. Venant (1870) and Levy (1870) represented such a giant step forward in continuum mechanics. For real-world applications, however, we needed further simplification and idealization including in some cases to ignore the relatively small elastic strain increments and consider only the plastic-strain increments and also to ignore the initial state of the material.

In the following, we use the customary subscript notations for rectangular Cartesian coordinates to introduce the subject of the conventional theory of plasticity. The key features of the plastic behavior of ductile metals to be captured in the mathematical theory of plasticity include irreversibility, time independence, very low stiffness of resistance to further plastic deformation when compared with its resistance to elastic deformation, and large ductility itself. The general form of plastic stress–strain relations to describe these key features accurately can be very complicated, and drastic idealizations of plastic behavior must be made before they can be placed on the most powerful computer for solutions of interesting problems. These idealizations and simplifications leading to the development of simple plastic theory are presented in the following including their applications to engineering design.

3.1.2 Basic Field Equations in Plasticity

Employing either the elasticity theory or the plasticity theory in continuum mechanics requires three basic sets of relations to be fulfilled in order to obtain a solution: (1) equilibrium conditions, (2) compatibility conditions, and (3) constitutive relations. The first two sets of relations are identical in both elasticity and plasticity. Only the constitutive relations are different in the two theories.

Since the plastic deformations are highly irreversible and load-path dependent, the constitutive relations for an elastic–plastic material should be written in incremental form as

$$d\sigma_{ij} = C_{ijkl}^{ep} d\varepsilon_{kl} \qquad (3.1)$$

where

$d\sigma_{ij}$ and $d\varepsilon_{kl}$ are stress and strain increment tensors, respectively

C_{ijkl}^{ep} is the response tensor or the tangent stiffness of the material

This tensor can be expressed as

$$C_{ijkl}^{ep} = C_{ijkl} + C_{ijkl}^{p} \qquad (3.2)$$

where

C_{ijkl} represents the elastic response of the material

C_{ijkl}^{p} accounts for the difference between the elastic–plastic response and the elastic response (plasticity influence)

The plasticity response tensor C_{ijkl}^{p} is a function of the current state of stress and strain and the yield function; that is, it is load-path dependent, as will be illustrated later.

Equations 3.1 and 3.2 provide the most general formulation of the constitutive relations for an elastic–plastic material. From these relations, it is clear that the stress increments $d\sigma_{ij}$ can be uniquely determined in terms of the yield function and the strain increments $d\varepsilon_{ij}$. However, the strain increments $d\varepsilon_{ij}$ cannot be defined in terms of the current state of stress and stress increments $d\sigma_{ij}$.

The different aspects of plasticity, yield criteria (or failure criteria), incremental strain, and constitutive relations are illustrated in the following subsections for an elastic–plastic material.

3.1.3 Yield Criteria Independent of Hydrostatic Pressure

The elastic limit of a material is defined as yielding and is determined under the combined state of stresses by a yield criterion. For a simple tension test, this limit is the yield stress, σ_o, while in shear test it is the yield shear stress, τ_o. For the general state of stress, this limit can be expressed as

$$f(\sigma_{ij}, k_1, k_2, \ldots) = 0 \qquad (3.3)$$

where k_1, k_2, ... are material constants to be determined experimentally.

For isotropic materials, the orientation of the principal stresses is immaterial, and the values of the three principal stresses suffice to describe the state of stress uniquely. A yield criterion therefore consists in a relation of the form

$$f(\sigma_1, \sigma_2, \sigma_3, k_1, k_2, \ldots) = 0 \tag{3.4}$$

Since the principal stresses can be expressed in terms of combinations of stress invariants, I_1, J_2, and J_3, where I_1 is the first invariant of the stress tensor σ_{ij}, ($I_1 = \sigma_{kk}$), and J_2 and J_3 are, respectively, the second and third invariants of deviatoric stress tensor s_{ij}, $s_{ij} = \sigma_{ij} - (1/3)\sigma_{kk}\delta_{ij}$. Thus, Equation 3.4 can be replaced with

$$f(I_1, J_2, J_3, k_1, k_2, \ldots) = 0 \tag{3.5}$$

For isotropic materials independent of hydrostatic pressure, such as metals, the influence of hydrostatic pressure on yielding is not appreciable; therefore, the yield criterion can be reduced to

$$f(J_2, J_3, k_1, k_2, \ldots) = 0 \tag{3.6}$$

This criterion can still be simplified further as in the von Mises and Tresca yield criteria.

The von Mises yield criterion states that yielding begins when the octahedral shearing stress reaches a critical value k:

$$\tau_{oct} = \sqrt{\frac{2}{3}J_2} = \sqrt{\frac{2}{3}}k \tag{3.7}$$

which reduces to the simple form

$$f(J_2) = J_2 - k^2 \tag{3.8}$$

or in terms of the principal stresses σ_1, σ_2, and σ_3

$$(\sigma_1 - \sigma_2)^2 + (\sigma_2 - \sigma_3)^2 + (\sigma_3 - \sigma_1)^2 = 6k^2 \tag{3.9}$$

The constant k is the yield stress in pure shear, which can be simply determined by applying Equation 3.9 to uniaxial test

$$k = \frac{\sigma_o}{\sqrt{3}} \tag{3.10}$$

where σ_o is the material yield stress. Equation 3.9 can be represented geometrically by an ellipse for the case of biaxial state of stress, Figure 3.1a, and by a cylinder in

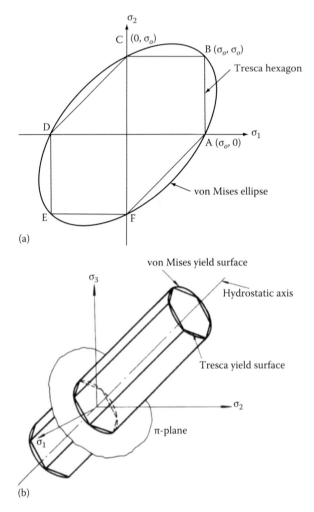

FIGURE 3.1 Geometric representation of von Mises and Tresca yield criteria: (a) yield criteria in the coordinate plane $\sigma_3 = 0$ and (b) yield surfaces in principal stress space.

the principal stress space as shown in Figure 3.1b. In relation with Equation 3.8, the von Mises criterion is often referred to as the J_2-theory. This equation also represents the simplest mathematical expression of the yield criterion for hydrostatic pressure–independent material, Equation 3.6.

The Tresca yield criterion states that yielding would occur when the maximum shearing stress at a point reaches a critical value k. In terms of principal stresses, the maximum shear stress is the maximum value of half the difference between the principal stresses taken in pairs; that is,

$$\max\left(\frac{1}{2}|\sigma_1 - \sigma_2|,\ \frac{1}{2}|\sigma_2 - \sigma_3|,\ \frac{1}{2}|\sigma_3 - \sigma_1|\right) = k \tag{3.11}$$

The material constant k can be determined from the simple tension test

$$k = \frac{\sigma_o}{2} \tag{3.12}$$

The Tresca yield criterion can be represented geometrically by a hexagon for the case of biaxial state of stress, Figure 3.1a, and by a regular hexagonal prism in the principal stress space, as shown in Figure 3.1b. In this criterion, yielding is governed by the maximum and minimum principal stresses, while in the von Mises criterion, the three principal stresses have a role in the yield criterion.

3.1.4 YIELD CRITERIA FOR PRESSURE-DEPENDENT MATERIALS

The yielding of most metals is hydrostatic pressure independent. On the other hand, the behavior of many nonmetallic materials, such as soil, rock, and concrete, is characterized by its hydrostatic pressure dependence. Therefore, the stress invariant I_1 should not be omitted from Equation 3.5. The well known are the Mohr–Coulomb and Drucker–Prager criteria, among others, which describe the failure surface of a pressure-dependent material.

Mohr's criterion is a generalized version of the Tresca criterion; in both criteria, the maximum shear stress is the only decisive measure of impending failure. However, while the Tresca criterion assumes that the critical value of the shear stress is a constant, Mohr's failure criterion considers the limiting shear stress τ in a plane to be a function of the normal stress σ in the same plane at a point; that is,

$$|\tau| = f(\sigma) \tag{3.13}$$

where $f(\sigma)$ is an experimentally determined function.

The simplest form of the Mohr envelope $f(\sigma)$ is a straight line as illustrated in Figure 3.2. The equation of this line, which is known as Coulomb's equation, is

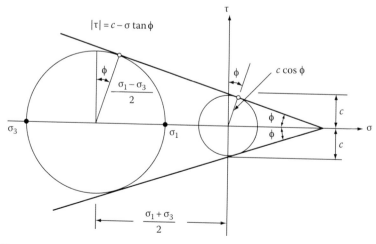

FIGURE 3.2 Mohr–Coulomb criterion with straight lines as the failure surface.

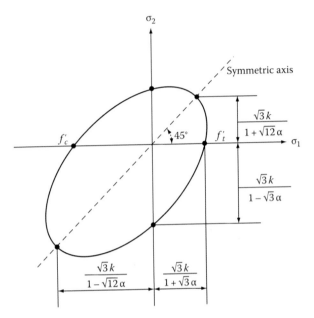

FIGURE 3.3   Drucker–Prager criterion in the coordinate plane $\sigma_3 = 0$.

$$|\tau| = c - \sigma \tan \phi \tag{3.14}$$

where
  c is the cohesion
  ϕ is the angle of internal friction

Both are material constants to be determined experimentally. The failure criterion associated with Equation 3.14 is referred to as the Mohr–Coulomb criterion. In the special case of frictionless materials, for which $\phi = 0$, Equation 3.14 reduces to the maximum-shear-stress criterion of Tresca, $\tau = c = k$.

The Drucker–Prager criterion is a simple modification of the von Mises criterion, where the influence of the hydrostatic stress component is introduced as follows:

$$f(I_1, J_2) = \alpha I_1 + \sqrt{J_2} - k = 0 \tag{3.15}$$

Equation 3.15 can be represented geometrically for the case of biaxial state of stress as an off-center ellipse, Figure 3.3, and in a principal stress space it is a right-circular cone. Both Mohr–Coulomb and Drucker–Prager criteria are shown in principal stress space in Figure 3.4.

3.1.5 INCREMENTAL STRAIN

It has been stated before that the constitutive relations of elastic–plastic materials have to be expressed in an incremental form. The total strain increment tensor can be

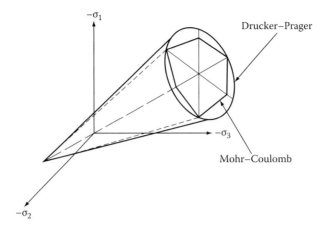

FIGURE 3.4 Mohr–Coulomb and Drucker–Prager criteria in principal stress space.

decomposed into two component tensors: one is the elastic strain increment tensor, $d\varepsilon_{ij}^{e}$, and the second is the plastic strain increment tensor, $d\varepsilon_{ij}^{p}$:

$$d\varepsilon_{ij} = d\varepsilon_{ij}^{e} + d\varepsilon_{ij}^{p} \tag{3.16}$$

The elastic strain increment tensor can be obtained from Hooke's law (Chapter 2):

$$d\varepsilon_{ij}^{e} = D_{ijkl}d\sigma_{kl} \tag{3.17}$$

or

$$d\varepsilon_{ij}^{e} = \frac{dI_1}{9K}\delta_{ij} + \frac{ds_{ij}}{2G} \tag{3.18}$$

where
  D_{ijkl} is the elastic compliance tensor, which is the inverse of C_{ijkl}
  I_1 is the first stress invariant
  K is the bulk modulus
  δ_{ij} is Kronecker's delta
  s_{ij} is the deviatoric stress tensor
  G is the shear modulus

Hence, the stress–strain relation for a plastic material reduces essentially to a relation involving the current state and the incremental changes of stress and plastic strain. This latter relationship is discussed in the following subsection.

3.1.6   Constitutive Equations for Perfectly Plastic Materials

Based on the von Mises potential function, the plastic flow equations can be written in the form

$$d\varepsilon_{ij}^p = d\lambda \frac{\partial g}{\partial \sigma_{ij}} \tag{3.19}$$

where $d\lambda$ is a scalar factor of proportionality, which is nonzero only when plastic deformations occur. The equation $g(\sigma_{ij}) = \text{constant}$ defines a surface (hyper-surface) of plastic potential in a nine-dimensional stress space. The direction cosines of the normal vector to this surface at the point σ_{ij} on the surface are proportional to the gradient $\partial g/\partial \sigma_{ij}$.

If the yield function and the plastic potential function coincide, $f = g$, then

$$d\varepsilon_{ij}^p = d\lambda \frac{\partial f}{\partial \sigma_{ij}} \tag{3.20}$$

and plastic flow develops along the normal to the yield surface $\partial f/\partial \sigma_{ij}$. Equation 3.20 is called the *associated flow rule* because the plastic flow is connected or associated with yield criterion, while Equation 3.19 with $f \neq g$ is called a nonassociated flow rule.

If the von Mises yield function is taken as the *plastic potential*

$$f(\sigma_{ij}) = J_2 - k^2 = 0 \tag{3.21}$$

Then, the flow rule has the simple form

$$d\varepsilon_{ij}^p = d\lambda \frac{\partial f}{\partial \sigma_{ij}} = d\lambda s_{ij} \tag{3.22}$$

where s_{ij} is the deviatoric stress tensor and the factor of proportionality $d\lambda$ has the value

$$d\lambda \begin{cases} = 0 & \text{wherever } f < 0 \text{ or } f = 0 \text{ but } df < 0 \\ > 0 & \text{wherever } f = 0 \text{ or } df = 0 \end{cases} \tag{3.23}$$

In the same manner, other flow rules can be associated with other yield functions such as Tresca and Mohr–Coulomb.

In order to determine the factor of proportionality, $d\lambda$, the consistency condition which ensures that the stress state $(\sigma_{ij} + d\sigma_{ij})$ existing after the incremental change $d\sigma_{ij}$ has taken place still satisfies the yield function f:

$$f(\sigma_{ij} + d\sigma_{ij}) = f(\sigma_{ij}) + df = f(\sigma_{ij}) \tag{3.24}$$

From this condition,

$$df = \frac{\partial f}{\partial \sigma_{ij}} d\sigma_{ij} = 0 \qquad (3.25)$$

Implementing Equation 3.22 in the relation of the stress increment tensor and elastic strain increment tensor gives

$$d\sigma_{ij} = C_{ijkl} d\varepsilon_{kl}^e = C_{ijkl} \left(d\varepsilon_{kl} - d\varepsilon_{kl}^p \right) = C_{ijkl} d\varepsilon_{kl} - C_{ijkl} d\lambda \frac{\partial f}{\partial \sigma_{ij}} \qquad (3.26)$$

Upon substituting Equation 3.26 into Equation 3.25

$$d\lambda = \frac{(\partial f/\partial \sigma_{ij}) C_{ijkl} d\varepsilon_{kl}}{(\partial f/\partial \sigma_{rs}) C_{rstu} (\partial f/\partial \sigma_{tu})} \qquad (3.27)$$

From Equations 3.26 and 3.27, the incremental stress–strain relation can be expressed explicitly as follows:

$$d\sigma_{ij} = \left[C_{ijkl} - \frac{C_{ijmn}(\partial f/\partial \sigma_{mn})(\partial f/\partial \sigma_{pq}) C_{pqkl}}{(\partial f/\partial \sigma_{rs}) C_{rstu} (\partial f/\partial \sigma_{tu})} \right] d\varepsilon_{kl} \qquad (3.28a)$$

that is, the elastic–plastic tensor of tangent moduli for an elastic-perfectly plastic material is

$$C_{ijkl}^{ep} = C_{ijkl} - \frac{C_{ijmn}(\partial f/\partial \sigma_{mn})(\partial f/\partial \sigma_{pq}) C_{pqkl}}{(\partial f/\partial \sigma_{rs}) C_{rstu} (\partial f/\partial \sigma_{tu})} \qquad (3.28b)$$

3.1.7 CONSTITUTIVE EQUATIONS FOR WORK-HARDENING MATERIALS

3.1.7.1 Material Hardening

The phenomenon whereby yield stress increases with further plastic straining is known as work-hardening or strain-hardening (Figure 3.5). The subsequent yield surface for an elastic–plastic material, which defines the boundary of the current elastic region, is the loading surface (Figure 3.5b). Thus, the loading surface may be generally expressed as a function of the current state of stress (or strain) and some hidden variables, expressed in terms of ε_{ij}^p and a hardening parameter k. Hence,

$$f(\sigma_{ij}, \varepsilon_{ij}^p, k) = 0 \qquad (3.29)$$

Determining the nature of the subsequent loading surfaces is one of the major problems in the work-hardening theory of plasticity. For a uniaxial behavior, the concepts

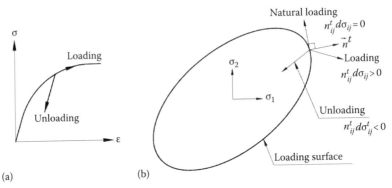

FIGURE 3.5 Loading criterion for a work-hardening material: (a) uniaxial case and (b) multiaxial case.

of "loading" and "unloading" are self-evident (Figure 3.5a); however, this is not the case under a multiaxial stress state. In order to define the modification of the subsequent yield surface during the process of plastic flow, there have been many rules of work-hardening. The most widely used rules are isotropic hardening, kinematic hardening, and a combination of both (mixed hardening).

For clarity, the general form of the loading function, Equation 3.29, can be written as

$$f\left(\sigma_{ij}, \varepsilon_{ij}^p, k\right) = F\left(\sigma_{ij}, \varepsilon_{ij}^p\right) - k^2(\varepsilon_p) = 0 \tag{3.30}$$

in which the hardening parameter k^2 represents the size of the yield surface, while the function $F(\sigma_{ij}, \varepsilon_{ij}^p)$ defines the shape of that surface. The parameter k^2 is expressed as a function of the effective plastic strain, ε_p, which is an integrated increasing function of the plastic strain increments but not the plastic strain itself:

$$\varepsilon_p = \sqrt{\frac{2}{3}\varepsilon_{ij}^p\varepsilon_{ij}^p} \tag{3.31}$$

For the case of uniaxial tension, ε_1^p reduces to ε_p. The value of ε_p depends on the loading history or the plastic strain path.

For a perfectly plastic material, the equation of the fixed yield surface has the form $F(\sigma_{ij}) = k^2$, where k is a constant. The simplest hardening rule, *isotropic hardening*, is based on the assumption that the initial yield surface expands uniformly without distortion or translation as plastic flow occurs (Figure 3.6). The size of the yield surface is now governed by the value k^2, which depends upon the plastic strain history. The equation for the subsequent yield surface or loading surface can be written in the general form

$$F(\sigma_{ij}) = k^2(\varepsilon_p) \tag{3.32}$$

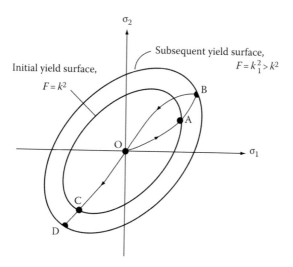

FIGURE 3.6 Subsequent yield surface for an isotropic-hardening material.

In the case of von Mises initial yield function, for example, $F=J_2$, Equation 3.32 becomes

$$J_2 = \frac{1}{2} s_{ij} s_{ij} = k^2(\varepsilon_p) \qquad (3.33)$$

When the effective stress $\sigma_e = \sqrt{3J_2}$ is introduced in Equation 3.33 as a hardening parameter, the isotropic-hardening von Mises model takes the form

$$f(\sigma_{ij},k) = \frac{3}{2} s_{ij} s_{ij} - \sigma_e^2(\varepsilon_p) \qquad (3.34)$$

where the hardening parameter $\sigma_e(\varepsilon_p)$ is related to the effective strain ε_p through an experimental uniaxial stress–strain curve.

On the other hand, in *kinematic hardening*, the loading surface is assumed to translate during plastic flow as a rigid body in stress space, while maintaining the size, shape, and orientation of the initial yield surface. As illustrated in Figure 3.7 as the point moves along its loading path from point A to point B, the yield surface translates (no rotation) as a rigid body. Thus, the subsequent yield surface will wind up in the position indicated in the figure when the stress point has reached position B. The new position of yield surface represents the most current yield function, whose center is denoted by α_{ij}. For unloading along the initial path of loading, path BAO, for example, the material behaves elastically from point B to point C and then begins to flow again before the stresses are completely relieved.

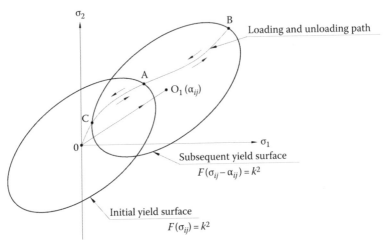

FIGURE 3.7 Subsequent yield surface for a kinematic-hardening material.

For kinematic hardening, the equation of loading surface has the general form

$$f(\sigma_{ij}, \varepsilon_{ij}^{p}) = F(\sigma_{ij} - \alpha_{ij}) - k^2 = 0 \tag{3.35}$$

where
  k is a constant
  α_{ij} are the coordinates of the center of the loading surface (or the vector OO_1 in Figure 3.7), which changes with plastic deformation

3.1.7.2 Incremental Plastic Strain

As discussed before, the general expression of a yield surface for work-hardening material has the form

$$f\left(\sigma_{ij}, \varepsilon_{ij}^{p}, k\right) = 0 \tag{3.36}$$

The plastic strain can be generally expressed by a nonassociated flow rule in the form

$$d\varepsilon_{ij}^{p} = d\lambda \frac{\partial g}{\partial \sigma_{ij}} \tag{3.37}$$

where $g = g(\sigma_{ij}, \varepsilon_{ij}^{p}, k)$, as for $f(\sigma_{ij}, \varepsilon_{ij}^{p}, k)$, is a known plastic potential as discussed before, and $d\lambda$ is a scalar function to be determined consistency condition, $df=0$.

  Following the same procedure as for perfectly plastic material, the following relation can be obtained (details in Chen and Han, 1988):

$$d\sigma_{ij} = C_{ijkl}^{ep} d\varepsilon_{kl} \tag{3.38a}$$

where

$$C_{ijkl}^{ep} = C_{ijkl} - \frac{1}{h} \frac{\partial f}{\partial \sigma_{mn}} C_{mnkl} C_{ijst} \frac{\partial g}{\partial \sigma_{st}} \qquad (3.38b)$$

where

$$h = -\frac{\partial f}{\partial \sigma_{ij}} C_{ijkl} \frac{\partial g}{\partial \sigma_{kl}} - \frac{\partial f}{\partial \varepsilon_{ij}^p} \frac{\partial g}{\partial \sigma_{ij}} - \frac{\partial f}{\partial k} \frac{dk}{d\varepsilon_p} \sqrt{\frac{2}{3} \frac{\partial g}{\partial \sigma_{ij}} \frac{\partial g}{\partial \sigma_{ij}}} \qquad (3.38c)$$

Equation 3.38 is valid only for the case of plastic loading; therefore, it is necessary to have a loading criterion in terms of the given strain increment instead of the stress increment since the stress increment is unknown. This criterion can be expressed as follows:

$$\frac{\partial f}{\partial \sigma_{mn}} C_{ijkl} d\varepsilon_{kl} \begin{cases} > 0, & \text{loading} \\ = 0, & \text{neutral loading} \\ < 0, & \text{unloading} \end{cases} \qquad (3.39)$$

For the case of neutral loading $d\varepsilon_{ij}^p = 0$ or $d\lambda = 0$, and for the case of unloading, the elastic stress–strain relationship $d\sigma_{ij} = C_{ijkl} d\varepsilon_{kl}$ should be used.

3.1.8 SOLUTION PROCESS: AN ILLUSTRATION

In order to illustrate some basic features and useful concepts of elastic–plastic deformation of a structure, the following example is presented. The example illustrates a thick-walled tube with closed ends under internal pressure, and, for simplicity, it is made of an elastic–perfectly plastic material. The tube cross section is shown in Figure 3.8 with an internal radius a and an external radius b. It is assumed

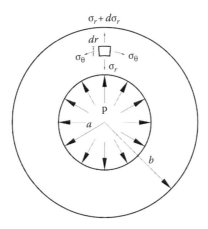

FIGURE 3.8 Cross section of a thin-walled tube under interior pressure.

that the tube is sufficiently long for end effects not to be felt in the zone under consideration.

Since the tube structure is axisymmetric, radial coordinates are more appropriate to use; the radial distance from the tube axis is r, the angular circumferential coordinate measured from an arbitrary datum is θ, and the axial distance from an arbitrary plane parallel to the radial axis is z.

3.1.8.1 Basic Equations

The only nontrivial *equilibrium* equation is

$$\frac{d\sigma_r}{dr} - \frac{\sigma_\theta - \sigma_r}{r} = 0 \tag{3.40}$$

With the assumption of a small displacement, if u is the radial displacement of a point at radius r, the *compatibility* equations are

$$\varepsilon_r = \frac{du}{dr} \tag{3.41}$$

and with the assumption of symmetrical deformation,

$$\varepsilon_\theta = \frac{u}{r} \tag{3.42}$$

$$\varepsilon_z = 0 \tag{3.43}$$

The compatibility relations are purely geometric and therefore they hold irrespective of whether the material is elastic or plastic.

The material tube is assumed elastic–perfectly plastic. In the elastic range, the behavior is described in terms of two elastic constants, Young's modulus E and Poisson's ratio ν. Since r, θ, and z are, by symmetry, the principal stress directions, the elastic constitutive relations are

$$E\varepsilon_r = \sigma_r - \nu(\sigma_\theta + \sigma_z)$$
$$E\varepsilon_\theta = \sigma_\theta - \nu(\sigma_r + \sigma_z) \tag{3.44}$$
$$E\varepsilon_z = \sigma_z - \nu(\sigma_r + \sigma_\theta)$$

Since $\varepsilon_z = 0$ (plane strain problem),

$$\sigma_z = \nu(\sigma_r + \sigma_\theta) \tag{3.45}$$

Thus, Equations 3.44 are modified to

$$E\varepsilon_r = (1-v^2)\sigma_r - v(1+v)\sigma_\theta$$

$$E\varepsilon_\theta = (1-v^2)\sigma_\theta - v(1+v)\sigma_r$$

(3.46)

The yield condition of Tresca is adopted, and the flow rule is associated with it by means of the normality condition.

The boundary conditions are

$$\sigma_r = 0 \quad \text{at } r = b$$

$$\sigma_r = -p \quad \text{at } r = a$$

(3.47)

where p is the interior pressure. In the axial direction, overall equilibrium requires

$$p\pi a^2 = \int_a^b 2\pi\sigma_z r dr$$

(3.48)

3.1.8.2 Elastic Solution

Upon elimination of u from Equations 3.41 and 3.42,

$$\varepsilon_r = \frac{d}{dr}(r\varepsilon_\theta)$$

(3.49)

After substituting ε_r and ε_θ in terms of σ_r and σ_θ from Equations 3.46 into Equation 3.49, and eliminating σ_θ with the aid of Equation 3.40, the following differential equation can be obtained:

$$\frac{d^2\sigma_r}{dr^2} + \frac{3}{r}\frac{d\sigma_r}{dr} = 0$$

(3.50)

The solution of Equation 3.50 is

$$\sigma_r = \frac{c_1}{r^2} + c_2$$

where c_1 and c_2 are the integration constants. Upon introducing the boundary conditions from Equation 3.47 and solving for c_1 and c_2, the following equation for σ_r can be obtained:

$$\sigma_r = \frac{pa^2(r^2-b^2)}{r^2(b^2-a^2)}$$

(3.51)

Substitution into Equation 3.40 gives

$$\sigma_\theta = \frac{pa^2(r^2 + b^2)}{r^2(b^2 - a^2)} \tag{3.52}$$

Equations 3.45 and 3.48 give a value of $v = 0.5$ and $\sigma_z = (1/2)(\sigma_r + \sigma_\theta)$.

The radial displacement is obtained from Equations 3.42 with the use of the second relation of (3.46):

$$u = r\varepsilon_\theta = \frac{1.5a^2b^2p}{E(b^2 - a^2)r} \tag{3.53}$$

The elastic stress distribution applies as long as the pressure p is small enough so that yield does not take place.

Since $\sigma_z = (1/2)(\sigma_r + \sigma_\theta)$, σ_z is the intermediate stress; that is,

$$\sigma_\theta > \sigma_z > \sigma_r \tag{3.54}$$

and hence the yield criterion of Tresca is

$$\sigma_\theta - \sigma_r = \sigma_o \tag{3.55}$$

where $\sigma_\theta - \sigma_r = \sigma_o$ is the yield stress in simple tension. Substitution of Equations 3.51 and 3.52 into (3.55) leads to

$$\sigma_\theta - \sigma_r = 2p\frac{b^2/r^2}{(b^2/a^2) - 1} = \sigma_o \tag{3.56}$$

From Equation 3.56, it is obvious that if the pressure is increased steadily, yield occurs at the inner surface first at a pressure

$$p = p_e = \frac{\sigma_o}{2}\left(1 - \frac{a^2}{b^2}\right) \tag{3.57}$$

which is a function of (b/a) and not the size of the tube.

3.1.8.3 Elastic–Plastic Expansion

If the internal pressure is increased beyond the value of the first yield (at inner surface), an enlarging plastic zone spreads outward from the inner surface; that is, the outer part of the tube remains elastic and the inner part becomes plastic (Figure 3.9).

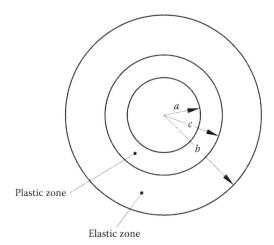

Plastic zone

Elastic zone

FIGURE 3.9   Plastic zone combined with an elastic zone of the thin-walled tube.

Suppose that the elastic–plastic boundary is at radius $r=c$ ($a \leq c \leq b$), as shown in Figure 3.9. For this boundary, the pressure from the inner part of the tube (plastic part) on the outer part, q, can be obtained from Equation 3.56 but with substituting for r and a with c,

$$q = \frac{\sigma_o}{2}\left(1 - \frac{c^2}{b^2}\right) \tag{3.58}$$

For the plastic zone, using Equation 3.55 to eliminate σ_θ from Equation 3.40, then integrating and introducing the boundary condition, $\sigma_r = -q$ at $r=c$, gives

$$\sigma_r = -q + \sigma_o \ln\frac{r}{c} \tag{3.59}$$

Substituting Equation 3.58 into (3.59) and (3.55) gives the stresses in the yield zone

$$\sigma_r = \sigma_o\left[\ln\frac{r}{c} - \frac{1}{2}\left(1 - \frac{c^2}{b^2}\right)\right]$$

$$\sigma_\theta = \sigma_o\left[\ln\frac{r}{c} + \frac{1}{2}\left(1 + \frac{c^2}{b^2}\right)\right] \tag{3.60}$$

Employing the boundary condition $\sigma_r = -p$ at $r=a$, we obtain

$$p = q + \sigma_o \ln\left(\frac{c}{a}\right) = \frac{\sigma_o}{2}\left(1 - \frac{c^2}{b^2}\right) + \sigma_o \ln\left(\frac{c}{a}\right) \tag{3.61}$$

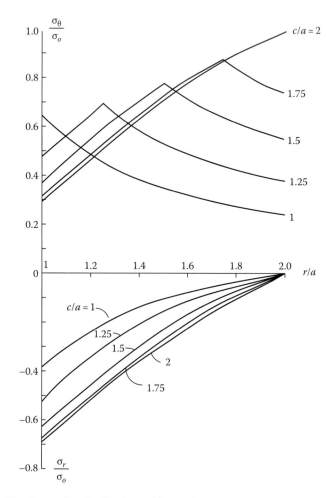

FIGURE 3.10 Successive distributions of circumferential and radial stresses in the elastic–plastic expansion of a tube; $b/a=2$.

Hence, for any value of c between a and b, the corresponding pressure may be calculated, and also σ_θ and σ_r are determined throughout the tube. Figure 3.10 shows the results of a tube with $b/a=2$ for different values of c/a.

Following the same thinking, the elastic–plastic deformation can be determined and the unloading behavior can be predicted as well.

3.2   LIMIT THEOREMS OF PERFECT PLASTICITY

3.2.1   Introduction

In order to carry out plastic analysis and design effectively in the real world of engineering, herein, we shall deal with idealizations of idealizations. As we have observed previously, once the material is well into the plastic range, it exhibits

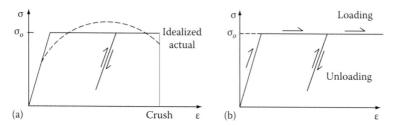

FIGURE 3.11 Uniaxial stress–strain relationship of an elastic–perfectly plastic material: (a) material with limited plastic strain and (b) highly ductile material.

relatively low additional resistance to increasing load. This feature can be simply captured by ignoring the small resistance in the plastic range and idealize the material as perfectly plastic. The consequence of such an idealization and the selection of proper flow strength depends on the problem to be solved as on the material itself. For example, for moderate plastic strain range in most structural engineering design, the flow strength to be chosen for the perfect plasticity idealization should be the average strength of the applicable range of the strains of the nonlinear stress–strain behavior as represented by the horizontal solid line in Figure 3.11a.

3.2.2 WHY LIMIT ANALYSIS?

In order to obtain a valid solution in continuum mechanics, three conditions should be satisfied: equilibrium, compatibility, and constitutive relations. For some cases, it is difficult to satisfy all three conditions. For simplicity, in limit analysis instead, only two conditions of the three are necessary for a simpler solution. Historically, engineers, based on intuition, developed many simple solutions for weak-tension materials (based on equilibrium) and for ductile materials (from kinematics), which have now been justified by limit analysis. The theorems of limit analysis give us a very powerful tool to estimate the upper and lower bounds of the collapse load of structures or structural members without having to go through a very tedious calculation procedure. In both theorems, strain-hardening of the material is ignored but its effect can be reflected realistically in the selection of a proper level of flow strength as illustrated in the preceding section, which is acceptable from a practical point of view. This further idealization of perfect plasticity enables the proof of the powerful limit theorems, which provide an excellent guide for preliminary design as well as analysis of structure. The development of the limit theorems and their illustrative engineering applications are described in the following.

In case of a lower-bound solution of the collapse load, only equilibrium and yield criterion are satisfied; equilibrium is satisfied for stress or generalized stress. The solution so obtained represents a good safe guidance for the structural engineer and can be used to verify solutions from other methods quickly. The method is useful for application to different materials especially that of a tension-weak material, for example, stones or concrete. Hence, the safety of monumental structures such as cathedrals can be checked very well with such a simple equilibrium method following the flow of forces using simple hand calculations.

In upper-bound solutions of the collapse load, only kinematics and yield criterion are satisfied. The method is especially good for ductile materials and even applicable to some materials with limited ductility but with modifications. The method uses the engineer's physical intuition on failure modes and their corresponding collapse analysis can be made by hand calculations. Thus, it gives the engineer enough clarity of his or her vision to produce a structure that is understandable and works well with the force of nature.

3.2.3 BASIC ASSUMPTIONS

The collapse load obtained from limit analysis is different from the actual plastic collapse load since it is calculated for an ideal structure in which the deformation is assumed to increase without limit while the load is held constant. Of course, this assumption is not expected to happen in real structures but only in idealized structures in which neither work-hardening of the material nor significant changes in geometry of the structure occur. However, the limit load still represents a good estimate of the real collapse load.

The idealization of a structure analyzed using the limit analysis theorems comes from the following two basic assumptions:

1. *Perfectly plastic material*: The material of the structure is assumed to be perfectly plastic with associated flow rule without strain-hardening or softening (Figure 3.11b). In this simplification, many effects are ignored; for instance, effect of time is eliminated from calculations; effect of residual stresses on initial yielding and effect of local buckling on maximum plastic moment capacity of steel sections are ignored. In addition, the complex states of stresses and strains in reinforced concrete as a result of bond and cracks are very much simplified.
2. *Small deformation of the structure*: The changes in geometry of the body or the structure, which may occur at limit load, are negligible; hence, the geometric description of the body or structure remains unchanged during the deformation at the limit load. This assumption allows the use of the principle of virtual work, which is the key to the proof of the limit theorems.

3.2.4 LOWER-BOUND THEOREM

This theorem states that "if an equilibrium distribution of stress can be found which balances the applied loads, and is everywhere below yield or at yield, the structure will not collapse or will be just at the point of collapse."

Hence, the lower-bound theorem requires the justification of only two of the three sets of conditions necessary for solution in continuum mechanics, that is, equilibrium and yield condition (material law). This theory therefore expresses the ability of an ideal body to adjust itself to carry the applied loads if at all possible. In practice, the application of the lower-bound theorem has different versions depending on the structural material of the system. For example, in steel frames the method is called the *statical method* while in concrete there is the *strut-and-tie model* (STM) method.

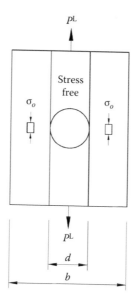

FIGURE 3.12 Lower-bound solution of a bar with a hole.

In order to illustrate the application of the lower-bound theorem, the following two examples are discussed. In the first example, Figure 3.12, a long prismatic bar of a rectangular cross section with one hole is subjected to an axial force P. If the yield stress of the bar material is σ_o, the simple two discontinuous stress fields shown in the figure can be assumed. In the two stress fields the bar is divided into strips, the continuous strips have simple tension σ_o and the discontinuous strips are stress free. Then, the lower-bound load of this bar is

$$P^L = \sigma_o(b-d)t \qquad (3.62)$$

The second example is a rigid punch indentation into a half-space of a perfectly plastic material (Figure 3.13). Assume that the width of punch in the direction perpendicular to the plane of paper is so large that this is a plane strain problem.

As a first attempt consider the simple discontinuous stress field shown in Figure 3.13a, which yields a lower bound on the limit load:

$$P_1^L = \sigma_o b = 2kb \qquad (3.63)$$

This is of course not a good lower bound because the load is considered to be carried only by a single vertical strip of material directly beneath the punch. To improve the answer, consider adding a horizontal pressure field as shown in Figure 3.13b. In the overlapping region, the material is subjected to a biaxial compression so that the vertical stress can be increased to $2\sigma_o$ without violating the yield condition. The improved lower bound obtained is

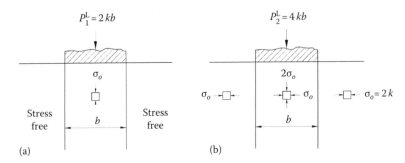

FIGURE 3.13 Stress fields for punch indentation in plane strain: (a) simple compression and (b) biaxial compression.

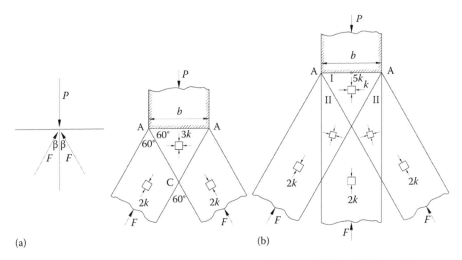

FIGURE 3.14 Load P carried by truss bars: (a) two-leg stress field and (b) three-leg stress field.

$$P_2^L = 2\sigma_o b = 4kb \tag{3.64}$$

Alternatively, the concept of truss-action approach can be assumed. The load P is carried by two inclined truss bars as shown in Figure 3.14a, and further, a vertical leg of amount $2k$ is added directly below the punch area AA to give the stress field shown in Figure 3.14b. In this case, the stress discontinuities are admissible. It is noted that the yield condition is violated in regions I and II. In region I, for example, the difference between the greatest and the least principal stress is $4k$. This violation can be accommodated by introducing at the free surface a horizontal strip in which there is a horizontal compressive stress $2k$. The width of this strip is as shown in Figure 3.15. Using this stress field, a better lower-bound solution can be obtained:

$$P_3^L = 5kb \tag{3.65}$$

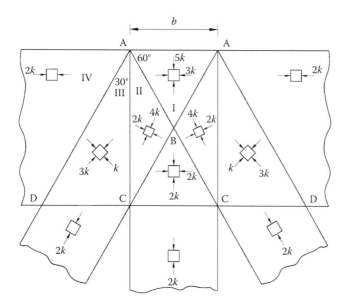

FIGURE 3.15 Combined stress field.

3.2.5 UPPER-BOUND THEOREM

This theorem states that "the structure will collapse if there is a compatible pattern of plastic failure mechanism for which the rate of work of the external forces equals or exceeds the rate of internal dissipation."

The upper-bound theorem thus requires the justification of only two of the three sets of conditions necessary for solution in continuum mechanics, that is, kinematics and yield criterion. The theory therefore states that if a path of failure exists, the ideal body will not stand up. In practice, the mechanism method of steel beams and frames and the yield line theory of concrete slabs are two different versions of applications of the upper-bound theorem.

In order to illustrate the application of the upper-bound theorem, the following two examples are discussed. The first example is the bar with one hole, Figure 3.12, which was solved using the lower-bound theorem. For this bar, three different compatible discontinuous failure modes are shown in Figure 3.16. In mode 1, Figure 3.16a, the upper and lower parts of the bar move as rigid bodies relative to each other by sliding along planes AB and CD perpendicular to the face of the bar and making an angle α as shown in the figure. If the relative tangential velocity of separation is $\dot{\delta}$, the velocity of separation is $\dot{\delta} \sin \alpha$ and the rate of external work is then $P_1^U \dot{\delta} \sin \alpha$. The rate of energy dissipation over the whole sliding surface is $k\dot{\delta}(b-d)t/\cos \alpha$. Hence,

$$P_1^U = \frac{2k(b-d)t}{\sin 2\alpha}$$

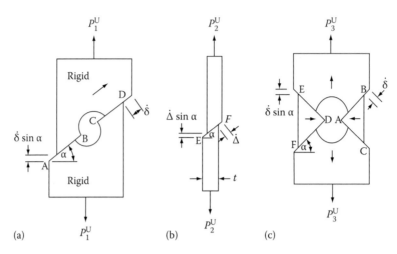

FIGURE 3.16 Kinematically admissible velocity fields of a bar with a hole: (a) in-plane shearing, mode 1, (b) out-of-plane shearing, mode 2, and (c) symmetric shearing, mode 3.

For a minimum value of P_1^U, sin 2α is set equal to 1 ($\alpha = 45°$). This gives

$$P_1^U = 2k(b - d)t \tag{3.66}$$

where k is the yield shear stress, which is equal to $(\sigma_o/\sqrt{3})$ according to von Mises and $(\sigma_o/2)$ according to Tresca. Based on the von Mises yield criterion,

$$P_1^U = \frac{\sigma_o}{\sqrt{3}}(b - d)t = 1.15 P_1^L \tag{3.67a}$$

Based on the Tresca yield criterion,

$$P_1^U = \frac{\sigma_o}{2}(b - d)t = P_1^L \tag{3.67b}$$

Mode 2, if the bar is assumed very thin (i.e., it is assumed plate), and mode 3 will give the same results as mode 1.

The second example is a rigid punch indentation into a half-space of a perfectly plastic material, Figure 3.17, for which a lower-bound solution was derived in the preceding section. Since the punch is assumed to be rigid, the geometric boundary condition requires that the movement of the contact plane must always remain plane. Two types of mechanisms, rotational and translational, are discussed in the following.

The simple rigid-body rotational mechanism about O, Figure 3.17b, is considered geometrically admissible if there are no constraints to hold the punch vertical. The block of material B rotates as a rigid body about O with an angular velocity $\dot{\alpha}$, and

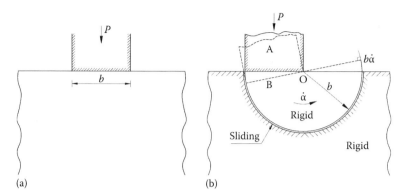

FIGURE 3.17   (a) Punch indentation problem and (b) rotational mechanism.

there is a semicircular transition layer between the rotating material and the remainder of the body. Since the angular velocity is $\dot{\alpha}$, the rate of work done by the external force P is the downward velocity at the center of the punch, $\dot{\alpha}b/2$, multiplied by P, while the total rate of energy dissipation is along the semicircular discontinuity surface and is found by multiplying the length of this discontinuity, πb, by the yield stress in pure shear, k, times the velocity across the surface $b\dot{\alpha}$. Equating the rate of external work to the rate of total internal energy dissipation gives

$$P^{\mathrm{U}}\left(\frac{1}{2}b\dot{\alpha}\right) = k(b\dot{\alpha})(\pi b)$$

or

$$P^{\mathrm{U}} = 2\pi kb = 6.28kb \tag{3.68}$$

It is noted that the upper-bound solution is independent of the magnitude of the angular velocity $\dot{\alpha}$, which means that $\dot{\alpha}$ can be assumed to be sufficiently small not to disturb the overall geometry. In other words, the proofs of the limit theorems can carry through using the initial geometry of the problem.

The rotational mechanism of Figure 3.17b may be generalized by taking the radius and the position of the center of the circle as two independent variables, aiming to obtain a better upper-bound estimate. If the center is shifted to O', as shown in Figure 3.18a, the rate of external work is $P^{\mathrm{U}}(r\cos\theta - b/2)\dot{\alpha}$, where r is the radius of the surface of discontinuity and θ is the angle between the face of the half-space and the line AO'. The rate of energy dissipation is given as $kr(\pi - 2\theta)r\dot{\alpha}$, and the resulting upper-bound solution is

$$P^{\mathrm{U}} = \frac{k(\pi - 2\theta)r^2}{r\cos\theta - (b/2)}$$

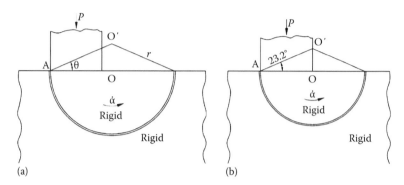

FIGURE 3.18 Rotating block with center at O′ for the purpose of minimizing the upper-bound estimate: (a) arbitrary rotational center and (b) optimal rotational center.

The obtained result may be optimized by minimization with respect to r and θ, which leads to the shape illustrated in Figure 3.18b and a better estimate of P^U,

$$P^U = 5.53kb \tag{3.69}$$

The mechanisms involving only rigid-body translations, shown in Figures 3.19 and 3.20, involve rigid-block sliding separated by plane velocity discontinuities. The mechanism of Figure 3.19a represents a rough punch, which requires the punch and the triangular block ABC have the same velocity and therefore move together. On the other hand, the mechanism of Figure 3.20a is referred to as a smooth punch, for it allows a relative sliding between the punch and the triangle ABC along the surface AB. It is noted that both the mechanisms are symmetrical about the center line and therefore only the right half of each is examined for kinematics.

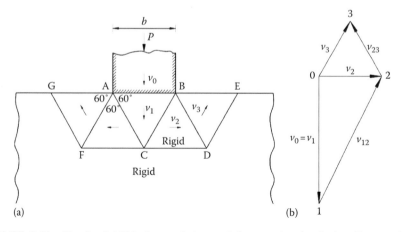

FIGURE 3.19 Simple rigid-block translation and the associated velocity diagram for a rough punch: (a) five-block mechanism and (b) velocity diagram.

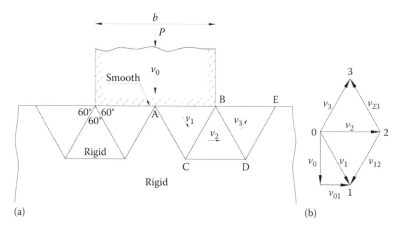

FIGURE 3.20 Rigid-block translation and the associated velocity diagram for a smooth punch: (a) six-block mechanism and (b) velocity diagram.

For a rough punch, the triangular region ABC in Figure 3.19a moves downward with a punch as a rigid body; that is, both have the same velocity, $v_1 = v_o$. However, for a smooth punch, the rigid block ABC would move with a different velocity v_1, Figure 3.20a, which has a horizontal component in addition to the vertical component v_o. The two triangular regions of the material BCD and BDE move as rigid bodies in the directions parallel to CD and DE, respectively. The velocity of the triangle BCD is determined by the condition that the relative velocity v_{12} between this triangle and the triangle in contact with punch must have the same direction BC. The velocity of the third triangle is determined in a similar manner. The information regarding velocities is represented by the velocity diagrams (or hodographs) as shown in Figures 3.19b and 3.20b.

From Figure 3.19b, the velocities of the rough punch mechanism of Figure 3.19a are

$$v_2 = v_3 = v_{23} = \frac{v_{12}}{2} = \frac{v_o}{\sqrt{3}}$$

and the work equation is

$$P^U v_o = 2k(bv_{12} + bv_2 + bv_{23} + bv_3)$$

Thus,

$$P^U = 5.78kb \tag{3.70}$$

From Figure 3.20b, the velocities of the smooth punch mechanism of Figure 3.20a are

$$v_1 = v_2 = v_3 = v_{12} = v_{23} = \frac{2v_o}{\sqrt{3}}$$

and the work equation is

$$P^U v_o = kb(v_1 + v_{12} + v_2 + v_{23} + v_3)$$

Thus,

$$P^U = 5.78kb \qquad\qquad (3.71)$$

which is the same result as the that of the rough punch. It is noted that the best upper-bound estimate is that of the mechanism in Figure 3.18b, $P^U = 5.53\,kb$, while the best lower-bound estimate from the preceding section was $P^L = 5.0\,kb$. The correct limit load for this problem is $P = 5.14\,kb$.

3.3 BAR ELEMENT AS A START

3.3.1 GENERALIZED STRESS–GENERALIZED STRAIN RELATION

In this section, we shall formulate the plastic stress–strain relations in terms of a bar element from which the parts of a framework are composed. The bar element described here is an element or segment of a beam.

3.3.1.1 Assumptions

In the development of the generalized stress–generalized strain relation of a bar element, it is assumed that the element cross section has one axis of symmetry, which is in the plane of loading. The plastic deformation due to shear forces is neglected and that due to normal forces can be neglected too if the magnitude of the axial force is small. Consequently, the simple plastic theory is concerned with the development of the relationship between the bending moment and the curvature of a bar element.

3.3.1.2 Material

The uniaxial stress–strain relationship of a bar element in tension and in compression is a bilinear relation, elastic up to the yield stress, σ_o, and then perfectly plastic, as shown in Figure 3.11b. From this relationship, the cross section in Figure 3.21a with different strain profiles in Figure 3.21b will have the stress distributions in Figure 3.21c.

3.3.1.3 Moment–Curvature Relation

The elastic–plastic behavior of a bar element can be typically illustrated in the form of its moment–curvature relation illustrated in Figure 3.22. The relation is elastic (part OA—state (i) in Figure 3.21) until the most stressed outer fibers attain a yield strain (stress), and then the corresponding sectional moment is the yield moment. When the bending moment is further increased, the curvature begins to increase with a higher rate along AB. This corresponds to the spread of the yield from the most strained outer fibers inward toward the neutral axis of the cross section (state (ii) in Figure 3.21), which is known as the plastification of the contained plastic flow.

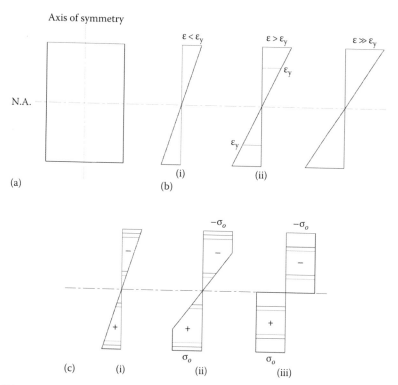

FIGURE 3.21 Section plastification: (a) cross section; (b) strain distribution; and (c) stress distribution.

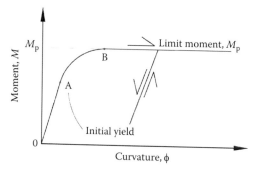

FIGURE 3.22 Typical moment–curvature relation.

Finally, the curvature tends to infinitely increase as the limiting value of the bending moment is approached (state (iii) in Figure 3.21). This limiting bending moment is called the fully *plastic moment*, M_p, or limit moment of the cross section.

The initial yield point in Figure 3.22 is defined by the *yield moment*, M_y, and the corresponding curvature, ϕ_y. The yield moment can be obtained from the elastic stress formula

$$M_y = \sigma_o Z_e \tag{3.72}$$

where Z_e is the elastic section modulus. For a rectangular section of thickness h and width b, $Z_e = bh^2/6$, and, therefore, $M_y = \sigma_o bh^2/6$. The yield curvature

$$\phi_y = \frac{\varepsilon_y}{y} \tag{3.73}$$

where y is the larger distance from the neutral axis to the most outer fibers. The slope of the elastic part OA can then be determined as (M_y/ϕ_y).

In this presentation, the effects of factors such as residual stresses in steel or localized crack or bond slip in reinforced concrete are ignored. Such a simplification affects the value of the yield moment; however, it does not affect the value of the plastic moment. Hence, for upper-bound or lower-bound solution, this simplification does not affect the collapse load estimates.

3.3.1.4 The Plastic Moment

For a bar cross section and segment in Figure 3.23, the plastic moment, M_p, can have a lower-bound value, M_p^L, and an upper-bound value, M_p^U, as follows. The lower-bound value can be obtained from the stress field in Figure 3.23d. The material above the plane of N.A. is in simple compression and below this plane, in simple tension, in both cases at yield σ_o. Equilibrium requires

$$\Sigma H = 0 \quad \text{or} \quad A_1 = A_2 \tag{3.74}$$

$$\Sigma M = 0 \quad \text{or} \quad M_p^L = \int_A \sigma_x y \, dA = \sigma_o \int_{A_1} y \, dA - \sigma_o \int_{A_2} y \, dA \tag{3.75}$$

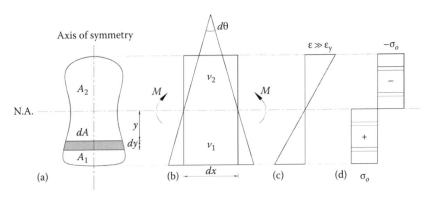

FIGURE 3.23   Section plastic moment: (a) cross section; (b) segment; (c) strain distribution; and (d) stress distribution.

In order to obtain the upper-bound value of plastic moment, M_p^U, the dissipated energy of the segment is

$$W_I = \int_V \sigma\varepsilon \, dV = \int_{V_1} \sigma_o \left(\frac{y \, d\theta}{dx} \right) dV + \int_{V_2} (-\sigma_o) \left(\frac{y \, d\theta}{dx} \right) dV = \sigma_o \left[\int_{A_1} y \, dA - \int_{A_2} y \, dA \right] d\theta$$

The external work is $W_E = M_p^U d\theta$, which upon equating with the dissipated energy gives

$$M_P^U = \sigma_o \int_{A_1} y \, dA - \sigma_o \int_{A_2} y \, dA \tag{3.76}$$

which is the same as M_p^L; thus,

$$M_p = \sigma_o \int_{A_1} y \, dA - \sigma_o \int_{A_2} y \, dA \tag{3.77}$$

For a rectangular section of thickness h and width b, $M_p = \sigma_o bh^2/4$ while $M_y = \sigma_o bh^2/6$. The ratio between the plastic moment, M_p, and the yield moment, M_y, for this section is 1.5. This ratio is called the shape factor, which varies for different cross-sectional shapes.

Tracing the generalized stress–generalized strain relation (moment–curvature relation) is thus a straightforward matter. By calculating M_y and ϕ_y the elastic part is determined. The elastic–plastic part is derived by assuming different values of curvature, ϕ, calculating the corresponding strain profile and hence the corresponding stress distribution. The moment corresponding to the assumed curvature can be calculated from equilibrium. As for the plastic part, which is not a real horizontal line, it can be approximated as a horizontal line with the plastic moment as a limiting value.

3.3.2 SIMPLE PLASTIC HINGE AS FURTHER SIMPLIFICATION

3.3.2.1 Idealization

A typical generalized stress–generalized strain relation (moment–curvature relation) has been illustrated in Figure 3.22 and its derivation has been covered in the preceding section. This relation is based on idealized stress–strain relationship of the material (elastic–perfectly plastic). It is illustrated again in Figure 3.24 with and without consideration of residual stresses. It is realized that residual stresses have no effect on the value of plastic moment, M_p.

The moment–curvature relation in Figure 3.24 can be further simplified as a bilinear relation, which consists of an elastic part and a perfectly plastic part (the horizontal line), as shown in the figure. The slope of the elastic part can be obtained as explained in the preceding section (M_y/ϕ_y) and the plastic part is determined by the

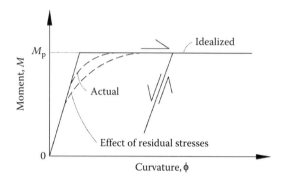

FIGURE 3.24   Generalized stress–generalized strain relation of a simple plastic hinge.

plastic moment, M_p. This idealized relation is derived from three quantities (M_y, ϕ_y, and M_p); that is, it can be obtained independently from the moment–curvature relation of the element.

Unloading from a plastic state to an elastic state is assumed to follow the path parallel to the elastic curve. As a result of this assumption for unloading, the elastic relationship between stress and strain is no longer unique in the sense that the behavior of the material may follow any elastic curve if the material unloads from a plastic state to an elastic state. Therefore, to perform an elastic–plastic analysis, it is necessary to follow an incremental procedure for a given history of loading in order to trace the unique states of moment and curvature in the cross section.

3.3.2.2   Concept

According to the bilinear idealization of the typical generalized stress–generalized strain relation of a bar element, Figure 3.24, a section attaining its plastic moment capacity undergoes plastic rotation without any further increase in bending moment. However, the state of bar plastification in the longitudinal direction is as illustrated in Figure 3.25, where plastic flow is contained except at the section of M_p. In other words, the section behaves like a real hinge while possessing a fully plastic moment. Thus, this idealized moment–curvature relation represents the generalized stress–generalized strain relation of what is known as "simple plastic hinge." This plastic-hinge behavior enables a structure to be analyzed continuously by inserting a plastic hinge at any section reaching its plastic moment.

FIGURE 3.25   Bar plastification in the longitudinal direction.

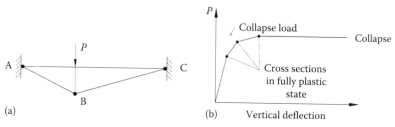

(a)

(b)   Vertical deflection

FIGURE 3.26 Elastic–plastic analysis of a beam example: (a) fixed-ended beam and (b) load–deflection relation.

In tracing the formation of the plastic hinges the structure becomes increasingly flexible until its stiffness is reduced to such a small value that imminent collapse occurs. In order to visualize the application of simple plastic hinge in the elastic–plastic analysis of framed systems, the beam of two fixed ends shown in Figure 3.26a is considered. The collapse mechanism of the beam requires the formation of three plastic hinges at the two ends A and C and under the point load, at B. The variation of the load P with vertical deflection at a point, point B for example, is plotted in Figure 3.26b, from which the following can be noted. The stiffness degradation of the beam as a result of the formation of plastic hinges is obvious and at collapse the structure stiffness is zero. The structure is elastic before the formation of any plastic hinges, plastic after the formation of the mechanism, and elastic–plastic between the two states. The behavior of a structure between the formations of any two consecutive plastic hinges is elastic and can be analyzed elastically.

3.3.2.3 Effect of Axial Force

The presence of axial force of large value noticeably reduces the plastic moment capacity of the bar element. For illustration, consider the rectangular section with the stress distributions shown in Figure 3.27. The stress distribution due to the bending moment, M, and normal force, N, is split into two parts: one due to M and the other due to N. From the figure,

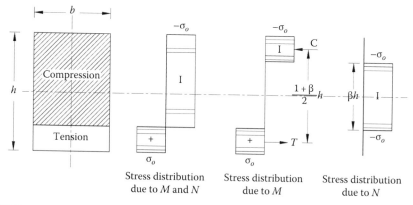

Stress distribution   Stress distribution   Stress distribution
due to M and N due to M due to N

FIGURE 3.27 Effect of axial force on section plastic moment.

$$N = \beta \sigma_o bh \quad \text{or} \quad \beta = \frac{N}{\sigma_o bh} = \frac{N}{N_p} \tag{3.78}$$

where N_p is the squash force of the section and hence the value of β can be considered as the magnitude of the axial force relative to the squash force, $\beta < 1.0$. Then, the total compressive force C, equal to the total tensile force T due to moment only is

$$C = T = \left(\frac{1-\beta}{2}\right)\sigma_o bh \tag{3.79}$$

Hence, the reduced plastic moment, M_{pr}, is

$$M_{pr} = C\left(\frac{1+\beta}{2}\right)h = \left(\frac{1-\beta^2}{4}\right)\sigma_o bh^2 \tag{3.80}$$

or

$$M_{pr} = (1-\beta^2)M_p \tag{3.81}$$

where M_p is the plastic moment of the section under pure bending. The relation in Equation 3.81 is derived for a rectangular section and other formulas can be derived for sections of different shapes.

In elastic–plastic analysis of structures, the consideration of the axial force in reducing the plastic moment capacity requires an iterative procedure and can be reasonably accounted for from a few iterations.

3.3.3 Elastic–Plastic Hinge-by-Hinge Analysis

In order to determine the collapse load of large-scale steel frames, the lower- or upper-bound methods, as illustrated in the subsequent subsections, may not be practical. An incremental elastic–plastic analysis, referred to as the *hinge-by-hinge method*, which is based on the plastic-hinge concept, is the appropriate tool. In this analysis (Wong, 2009), the load–deformation relation can be traced for increasing proportional loading until collapse, thus giving, in addition to the collapse load, a prediction of the deflection values at different loading levels.

The method consists of a series of elastic analyses, each of which represents the formation of a plastic hinge in the structure. The results for each elastic analysis are transferred to a spreadsheet from which the location for the formation of a plastic hinge and the corresponding load increment can be obtained. The method starts with the linear elastic stiffness matrix of the structure, and under proportional loading, the first plastic hinge to form is determined from moment checks and hence the first load increment is calculated. For the next load increment, the stiffness matrix is modified in order to account for the formation of the first plastic hinge and the analysis is performed to determine the second plastic hinge and the corresponding load factor. The procedure is repeated in this manner until the formation of the collapse mechanism and hence the final load increment and the collapse load are determined (Figure 3.26).

If the equilibrium equations, from which the structure stiffness matrix is derived, are written for undeformed geometry, the analysis is a first-order elastic–plastic analysis. On the other hand, if these equations are written for deformed geometry or the slope-deflection equations of a beam-column are used in the formulation, the analysis is a second-order elastic–plastic analysis.

3.3.4 PLASTIC-HINGE ANALYSIS BY EQUILIBRIUM METHOD

The plastic-hinge concept can be used to obtain either a lower-bound or an upper-bound estimate of the collapse load of framed systems. The lower-bound solution is obtained from the equilibrium method, which is discussed in this section. On the other hand, the upper-bound solution is obtained from the mechanism method, which will be discussed in the following section. In addition, the concept can be used to perform an elastic–plastic analysis of framed systems, which will be discussed in a subsequent section.

In the equilibrium method (or lower-bound theorem) the solution procedure is simplified since it is required to satisfy only two of the three sets of conditions necessary for solution in continuum mechanics, that is, equilibrium and yield criterion. The solution starts with assuming a suitable bending moment diagram of the structure, which satisfies both equilibrium and the yield criterion ($M \leq M_p$), formally called *admissible stress field*. Since the structure is statically indeterminate, the moment diagram can be easily constructed by superposing the free bending moment diagram (from a determinate system) upon reactant bending moment diagram(s) (from redundant force(s)). Different moment diagrams lead to different lower-bound estimates of the collapse load. The highest value of these estimates is the best lower-bound solution. It is noted that the bending moment diagram leading to the best estimate of collapse load will lead to the best collapse mechanism. The solution procedure using this method is illustrated by the following examples.

In the first example, the beam in Figure 3.28a is once statically indeterminate and the free moment and redundant moment diagrams are illustrated in Figure 3.28b and c, respectively, where the redundant is the moment at D, M_D. Upon superposing the two diagrams the final bending moment diagram is obtained as shown in Figure 3.28d. For the beam to reach the collapse load two plastic hinges should form. There are three possibilities for this: to have plastic hinges at B and C, at C and D, or at B and D. The three cases are examined in the following.

3.3.4.1 Case 1: Plastic Hinges at B and C

For this case and with reference to Figure 3.28d, upon equating the moments at B and C with M_p and solving for P_1^L and M_D, it is found that $P_1^L = M_p/L$ and $M_D = 0.5M_p$. This obtained value of M_D means that the yield condition is satisfied and hence this solution is valid.

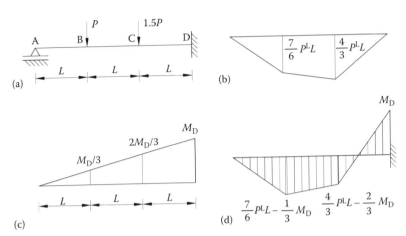

FIGURE 3.28 Lower-bound solution of a beam example: (a) beam; (b) free moment; (c) redundant moment; and (d) final moment.

3.3.4.2 Case 2: Plastic Hinges at C and D

For this case and with reference to Figure 3.28d, upon equating the moments at C and D with M_p and solving for P_2^L, it is found that $P_2^L = 1.25M_p/L$. When checking the value of the moment at B it is found that $M_B = 1.125M_p$, that is, the yield criterion is violated. For this solution to be valid, the collapse load has to be lowered in order to have $M_B = M_p$. Hence, the valid collapse load for this case is $P_2^L = (1.25M_p/L) \div 1.125 = 1.11M_p/L$, and of course the moments at C and D should be adjusted subsequently.

3.3.4.3 Case 3: Plastic Hinges at B and D

Upon following the previous procedure and equating the moments at B and D with M_p and solving for P_3^L, it is found that $P_3^L = 1.14M_p/L$. When checking the value of the moment at C it is found that $M_C = 0.857M_p$; that is, the yield criterion is satisfied and this solution is valid. From the solution of the three cases it is found that case 3 gives the highest value. Therefore, the best lower-bound solution is $P^L = 1.14M_p/L$.

The second example is the portal frame shown in Figure 3.29a. For simplicity the final moment is split into two diagrams: one due to the vertical load, Figure 3.29b, and the other due to the lateral load, Figure 3.29c. By superposing the two diagrams and obtaining the final diagram, Figure 3.29d, it is realized that the moments at C, D, and E can be equated with M_p. Hence, from equilibrium, the reactions at A can be calculated leading to $M_A = 5M_p - 10P^LL$ and $M_B = -3M_p + 6P^LL$. To determine P^L, assume M_A and check M_B and vice versa; values of M_p, $-M_p$, and 0 can be used for either M_A or M_B. From the obtained valid results it is found that the best lower-bound solution is $P^L = 0.6M_p/L$.

3.3.5 Plastic-Hinge Analysis by Mechanism Method

In the mechanism method (or upper-bound theorem), the solution procedure is simplified since it is required to satisfy only two of the three sets of conditions necessary

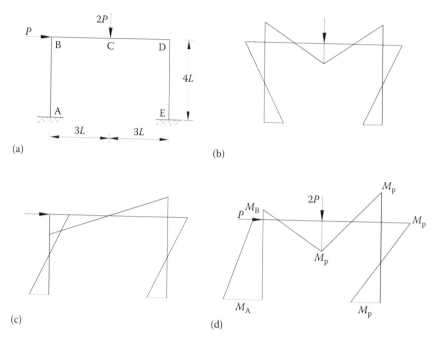

FIGURE 3.29 Lower-bound solution of a portal frame example: (a) frame; (b) assumed moment due to vertical load; (c) assumed moment due to lateral load; and (d) assumed final moment.

for solution in continuum mechanics, that is, kinematics and yield criterion. The solution starts with identifying all possible collapse mechanisms; then, the virtual work equation for each mechanism is established and the collapse load is obtained. The best upper-bound solution is the lowest estimate of the collapse loads of all mechanisms. The method is illustrated by the following examples.

The first example is the beam shown in Figure 3.30, which was solved for lower-bound collapse load, and the three possible mechanisms of the beam are illustrated in the figure. For the first mechanism, Figure 3.30b, the dissipated energy is $\{M_p\,(2\theta+3\theta)=5M_p\theta\}$ and the external work is $\{P_1^U(L\theta)+1.5P_1^U(2L\theta) = 4P_1^U L\theta\}$. Upon equating the dissipated energy with the external work, the collapse load for this mechanism is $P_1^U = 1.25M_p/L$. For the second mechanism, Figure 3.30c, the dissipated energy is $\{M_p(\theta+3\theta)=4M_p\theta\}$ and the external work is $\{P_2^U(2L\theta)+1.5P_2^U(L\theta) = 3.5P_2^U L\theta\}$, thus, the collapse load is $P_2^U = 1.14M_p/L$. The collapse load of the third mechanism is $P_3^U = 3.0M_p/L$. From the obtained results the best upper-bound solution is $P^U=1.14M_p/L$, which is equal to the lower-bound solution, which means that the obtained solution is the unique solution. It is noted that the best moment diagram of the best lower-bound solution and the best mechanism are the same.

The second example is the portal frame shown in Figure 3.31a, which was solved for lower-bound collapse load, and the three possible mechanisms are illustrated in the figure. For the first mechanism (the beam mechanism), Figure 3.31b, the

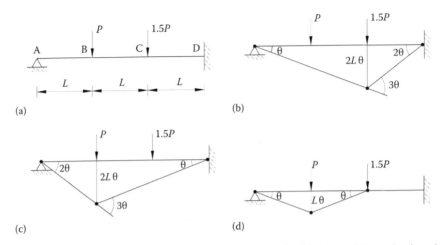

FIGURE 3.30 Upper-bound solution of a beam example: (a) beam; (b) mechanism 1; (c) mechanism 2; and (d) mechanism 3.

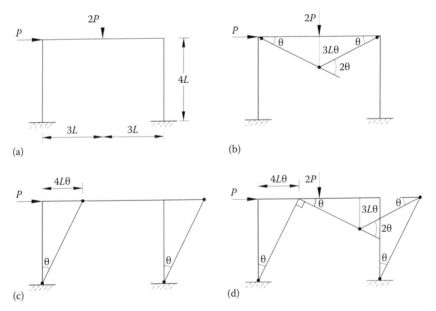

FIGURE 3.31 Upper-bound solution of a portal frame example: (a) frame; (b) beam mechanism (mechanism 1); (c) sway mechanism (mechanism 2); and (d) combined mechanism (mechanism 3).

dissipated energy is $4M_p\theta$ and the external work is $6P_1^U L\theta$; hence, the collapse load for this mechanism is $P_1^U = 0.67M_p/L$. For the second mechanism (the sway mechanism), Figure 3.31c, the collapse load is calculated from $M_p(4L\theta) = P_2^U(4L\theta)$, giving $P_2^U = M_p/L$. For the third mechanism, Figure 3.31d, which is a combination of the beam and sway mechanism, the dissipated energy is $M_p(6L\theta)$ and the external work is $P_3^U (10L\theta)$; hence, the collapse load for this mechanism is $P_3^U = 0.6M_p/L$. From the obtained results, the best upper-bound solution is that of the third mechanism, $P^U = 0.6M_p/L$, which is equal to the lower-bound solution; that is, the obtained solution is the unique solution. It is noted that the best moment diagram of the best lower-bound solution and the best mechanism are the same.

In the solution of the previous examples, it was assumed that the plastic moment of a member is a constant value and does not change because of the normal force. This simplification is acceptable for small normal force; otherwise, an iterative procedure should be adopted in order to account for the effect of normal force on plastic moment. It should also be noted that in the case of local mechanism there is no guarantee that the yield criterion is satisfied in the non-collapse part, and, therefore, it has to be checked in those regions.

With the introduction of the plastic-hinge concept, the simple plastic methods of structural analysis and design as described herein were rapidly developed in the 1960s for the purpose of calculating the plastic collapse loads of frame structures. It was a remarkable achievement that it was possible to calculate the collapse load directly without considering the intervening elastic–plastic range. Thanks to the simple plastic-hinge concept, the plastic methods of analysis may now be said to be fully developed as a sequence of elastic analysis. The plastic methods of analysis have since become an important part of the modern theory of structures for practicing structural engineers. It has become an integral part of our undergraduate teaching curriculum.

3.3.6   Refined Plastic Hinge toward Advanced Analysis

The simple plastic hinge used in Section 3.3.2 may be called *concentrated plastic hinge* where the zero-length plastic hinge can form suddenly from the limit of elasticity of a member. This concentrated plastic-hinge concept accounts for inelasticity but not the spread of yielding or plasticity at sections, or the influence of the residual stresses.

Depending on the geometry used to form the equilibrium equations, the elastic-plastic hinge method may be divided into *first-order* and *second-order* plastic analyses. For the first-order elastic–plastic analysis, the undeformed geometry is used, and nonlinear geometry effects are neglected. As a result, the predicted collapse is the same as the simple plastic analysis or rigid plastic analysis. In the second-order elastic–plastic-hinge analysis, the deformed shape is considered, and geometry nonlinearities can be included with the use of stability functions for a beam-column element to capture the second-order effect. Load–displacement curves for various analysis methods are shown in Figure 3.32. With reference to this figure, these methods are

1. First-order linear elastic analysis
2. Elastic buckling analysis
3. Second-order elastic analysis
4. Rigid plastic load analysis (lower-bound and upper-bound solutions)

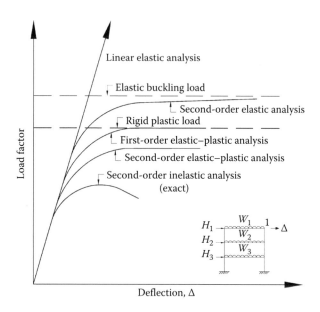

FIGURE 3.32   Load–deflection relation from various analysis methods.

5. First-order elastic–plastic analysis
6. Second-order elastic–plastic analysis
7. Second-order inelastic analysis (advanced analysis)

Desktop computing has now made it possible to combine the theory of stability and the theory of plasticity and apply them directly to frame design. This is known as the *second-order inelastic analysis* for frame design, or simply, the *advanced analysis* for frame design. The 2005 AISC LRFD specifications were issued to permit the use of this new analysis procedure for structural design.

Various methods have been proposed in literature in order to make this advanced analysis applicable to engineering practice. Among them, the *notational-load plastic hinge* is one of the most widely known. The notional-load plastic-hinge method is achieved by applying additional fictitious equivalent lateral loads to account for the influences of residual stresses, member imperfections, and distributed plasticity that are not included in the elastic-plastic-hinge method (Liew et al., 1994). With certain modifications, this method is accepted in the European Convention for Constructional Steelwork (ECCS, 1991).

To account for the spread of plasticity, a refined plastic-hinge method has been proposed to improve the accuracy of the notional-load plastic-hinge method. In the United States, to meet the current specification requirements, further modifications and calibration of the refined plastic hinge are made against the current AISC LRFD codes. A simple, concise, and reasonably comprehensive introduction to some of the advanced analysis methods developed in recent years is presented in Chapter 6. Details of this development can be found in Chen and Kim (1997).

3.4 CONCRETE PLATE ELEMENT AS A NEXT STEP

3.4.1 GENERALIZED STRESS–GENERALIZED STRAIN RELATION

Here, as in the bar element, the rectangular plate element is obtained by cutting through the entire thickness of a concrete slab. The relations between the values of bending moments and their respective rotations or curvatures for the ends of the sections represent the material behavior of the element. In what follows it will be necessary to define what is meant to be the state of initial yield, subsequent yield, and plastic collapse of the element in a two-dimensional space and of the structure as a whole. As we mentioned previously, the theories of reinforced concrete design do not deal with real reinforced concrete. We operate instead with an ideal composite material consisting of concrete and steel, the design properties of which have been approximated from those of real reinforced concrete by a process of drastic idealization and simplification.

3.4.1.1 Typical Moment–Curvature Relation

Reinforced concrete plates usually have light reinforcement and low shear stresses. The flexural behavior of such a structural member is typically illustrated in Figure 3.33 (*generalized stress–strain relation*). The behavior is linear until cracking takes place at a relatively small moment, point A. Then, the plate flexural stiffness is reduced; however, the behavior of cracked section is almost linear until the tension reinforcement yields, point B. At this point, the stiffness is greatly reduced and the change in curvature becomes relatively very large for a slight increase in the bending moment until failure takes place at a very large curvature, point C (nearly *elastic–perfectly plastic relation*).

3.4.1.2 Assumptions and Idealization

In the development of the generalized stress–generalized strain relation (or moment–curvature relation) of reinforced concrete plates, it is assumed that the reinforcement ratio is small (a fraction of the balanced reinforcement ratio) and the shear stresses are very low. Hence, the plate exhibits pure flexural behavior and very high ductility.

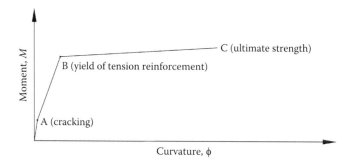

FIGURE 3.33   Typical moment–curvature relation.

It is also assumed that the tension reinforcement has an elastic–perfectly plastic stress–strain relationship, that is, no strain-hardening.

The generalized stress–generalized strain relation (or the moment–curvature relation) of a reinforced concrete plate can be idealized as a trilinear curve. The curve is defined by three generalized stresses (cracking moment, yield moment, and ultimate moment at points A, B, and C, respectively) and their corresponding generalized strains (curvatures).

In reality, of course, the stresses and strains in the plate element are much more complicated. As the loading increases, the concrete in the tension side of the segment starts to crack, progressively opens up, and moves upward diminishing the compression zone. When the loading increases further, more cracks develop, while the bonding between the reinforcing steel and its surrounding concrete starts to deteriorate. Finally, the reinforcement starts to yield resulting in an almost constant moment capacity with nearly an unlimited rotational deformation. Thanks to this approach, the complex local stress and strain states in the element are avoided and the field of application of the theory of plasticity to reinforced concrete structures can be realized and implemented in practice.

3.4.2 YIELD LINE THEORY AS A LOGICAL EXTENSION OF PLASTIC-HINGE ANALYSIS

3.4.2.1 Concept

Under overload conditions of a plate in flexure, the reinforcement will yield in a region of high moment. Consequently, this portion of the plate acts as a plastic hinge, with its ability to rotate but without any appreciable increase in its moment capacity. When the load is increased further, the hinging region rotates plastically, and the moment due to additional loads is redistributed to adjacent sections causing them to yield as shown in Figure 3.34. The bands in which yielding has occurred are referred to as yield lines, which divide the slab into a series of elastic plates. Eventually, enough yield lines exist to form a plastic mechanism in which the slab deforms plastically without an increase in the applied load.

Upon writing the equation of virtual work for this formed mechanism, an upper-bound estimate of the collapse load can be obtained. In this solution, only two sets of conditions are satisfied: kinematics and yield criterion.

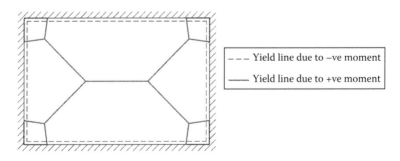

FIGURE 3.34 Yield lines in a plate fixed at four edges.

3.4.2.2 The Plastic Moment, M_p

The plastic moment (or nominal moment) of a slab cross section with a reinforcement A_s per unit length can be determined as follows (Figure 3.35). The compression stress resultant, C, is given as

$$C = 0.85 f_c' ab = 0.85 f_c' a \qquad (3.82)$$

where

f_c' is the concrete cylinder strength
a is the height of the compression stress block
b is the section width, which is equal to unit length

The tension in reinforcement, T is

$$T = A_s \sigma_o \qquad (3.83)$$

where σ_o is the reinforcement yield stress. Equilibrium in the horizontal direction leads to

$$C = T$$

or

$$0.85 f_c' a = A_s \sigma_o$$

This gives

$$a = \frac{A_s \sigma_o}{0.85 f_c'} \qquad (3.84)$$

Moment equilibrium leads to

$$M_p = M_n = T\left(d - \frac{a}{2}\right) \qquad (3.85)$$

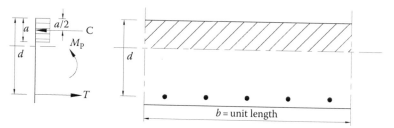

FIGURE 3.35 Equilibrium of a plate cross section.

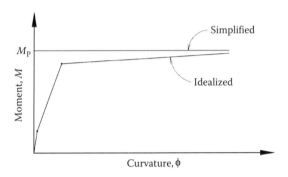

FIGURE 3.36　Yield criterion of a reinforced concrete plate.

Substituting Equations 3.83 and 3.84 into Equation 3.85 leads to

$$M_p = A_s \sigma_o \left(d - \frac{A_s \sigma_o}{1.7 f_c'} \right) \tag{3.86}$$

3.4.2.3　Yield Criterion

The idealized moment–curvature relation of a plate section with bending moment uniformly distributed along the section width has been discussed in the preceding section. This relation is illustrated by the trilinear curve in Figure 3.36. In the general case of a plate, the bending moment reaches its limit value at the most stressed regions first. Nevertheless, no rotation takes place until the moment spreads along a line, causing the two plates connected to this line to rotate.

Based on the deformation sequence of reinforced concrete slabs, the moment–curvature relation of a slab section can be further simplified by assuming a rigid plastic model. In this model, it is assumed that the slab does not exhibit elastic deformation and the limiting strength of the model is the plastic moment (Figure 3.36). This perfectly plastic behavior of the slab does not deviate much from reality since strain-hardening of reinforcement reduces the gap between the strength of the rigid model and that of the idealized model. In addition, elastic deformation is very small in comparison with measured or calculated maximum deformation.

3.4.3　YIELD-LINE ANALYSIS FOR CONCRETE SLAB DESIGN

3.4.3.1　Yield Criterion of Yield Line

Consider that the orthogonal reinforcement of a slab lays in the x- and y-directions as shown in Figure 3.37a. If yielding occurs along a line at an angle α to the x-direction, as shown in Figure 3.37b, the bending and twisting moments are assumed to be uniformly distributed along the yield line and are the maximum values provided by the flexural capacities of the reinforcement layers crossed by the yield line. The plastic moments as a result of the reinforcement in the x- and y-directions are M_{px} and M_{py} per unit length. The bending moment M_{pb} and the twisting moment M_{pt} per unit length of the yield line in Figure 3.37b can be calculated from the moment equilibrium of

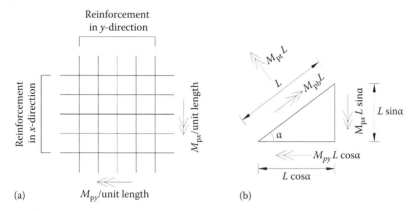

FIGURE 3.37 Yield criterion of a yield line: (a) reinforcement pattern and (b) moments of an element.

the element (MacGregor and Wight, 2005). The angle α is measured counterclockwise from the x-axis, and the bending moments M_{px}, M_{py}, and M_{pb} are positive if they cause tension in the bottom of the slab and the twisting moment M_{pt} is positive if its vector points away from the section as shown in the figure. From equilibrium,

$$M_{pb}L = M_{px}L\sin^2\alpha + M_{py}L\cos^2\alpha$$

or

$$M_{pb} = M_{px}\sin^2\alpha + M_{py}\cos^2\alpha \qquad (3.87)$$

and

$$M_{pt}L = M_{px}L\sin\alpha\cos\alpha - M_{py}L\sin\alpha\cos\alpha$$

or

$$M_{pt} = \frac{(M_{px} - M_{py})\sin 2\alpha}{2} \qquad (3.88)$$

In case the orthogonal reinforcement results in $M_{px} = M_{py}$, Equations 3.87 and 3.88 will be reduced to $M_{pb} = M_{px} = M_{py}$ and $M_{pt} = 0$ regardless of the value of the angle of the yield line.

3.4.3.2 Axes of Rotations and Yield Lines

As mentioned before, yield lines form regions of maximum moment and divide the plate into a series of elastic plate segments. When the yield lines have formed, any

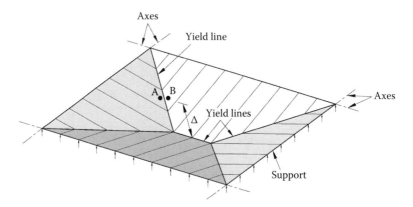

FIGURE 3.38   Deformations of a slab with yield lines.

additional deformation concentrates in yield lines, and the slab deflects as a series of stiff plates joined together by long hinges, as shown in Figure 3.38. The pattern of deformation is controlled by axes that pass along line supports and over columns, as shown in Figure 3.39, and by the yield lines. Since the individual plates rotate about the axes and/or yield lines, these lines must be straight. To satisfy the compatibility

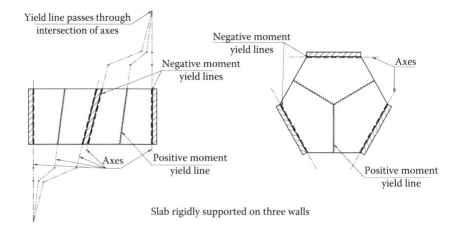

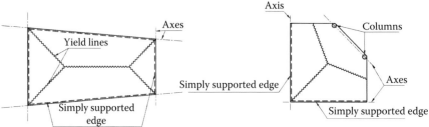

FIGURE 3.39   Examples of yield line patterns.

of deformations at points such as A and B in Figure 3.38, the yield line dividing two plates must intersect the intersection of the axes about which those plates are rotating. Figure 3.39 shows the locations of axes and yield lines in a number of slabs subjected to uniform loads.

3.4.3.3 Application

The application of the yield line theory to reinforced concrete slab is illustrated by the following three examples. The *first example* is the simply supported triangular slab shown in Figure 3.40a. It has a thickness of 200 mm, concrete strength $f_c' = 30$ MPa, and is reinforced with a bottom mesh of bars of diameter 16 mm every 125 mm in the *x*- and *y*-directions. The steel yield stress is $\sigma_o = 360$ MPa. It is required to calculate an upper-bound estimate of a uniformly distributed collapse load.

For simplicity it is assumed that the effective depth of reinforcement in both the *x*- and *y*-directions is 165 mm, which gives a plastic moment in both directions, 88.7 kN m/m. In this case, the plastic bending moment at any section will be 88.7 kN m/m and the twisting moment will be zero. The yield lines of the slab are shown with dashed lines for the positive moment, which divide the slab into three symmetric plates. Assume a vertical displacement δ at the slab center O. For the yield line AO the angles of rotation, Figure 3.40b, are

$$\theta_1 = \theta_1 = \frac{\delta}{2.667}$$

The total rotation of the positive moment at the yield line AO is $(\theta_1 + \theta_2)$ and the total dissipated energy for the three lines is

$$\left\{ M_p^{+ve} * 2 * \left(\frac{\delta}{2.667} \right) * 4.619 \right\} * 3 = 921.722\delta$$

Denoting the collapse load w_u^L, the external work of the three plates is

$$\left\{ w_u^L * \frac{2.309 \times 8}{2} * \left(\frac{\delta}{3} \right) \right\} * 3 = 9.236\delta w_u^L$$

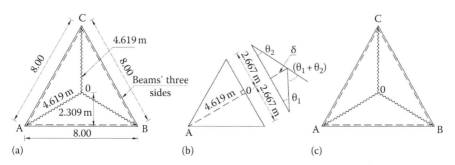

(a)   (b)   (c)

FIGURE 3.40 Triangular slab example.

Upon equating the dissipated energy and the external work,

$$w_u^L = 99.8 \, \text{kN/m}^2$$

The *second example* is the same slab in the previous example but with top reinforcement of bars of diameter 16 mm every 125 mm perpendicular to the slab edge in order to allow for continuity with adjacent panels. The negative moment in this case will be 88.7 kN m/m about a section parallel to the edge beams. The collapse mechanism in this case will be the same as before but with additional yield lines adjacent to the edge beams, as shown in Figure 3.40c. The angle of rotation at this negative moment lines will be $\theta_2 = \delta/2.309$. The total dissipated energy for the six lines is

$$\left\{ M_p^{+ve} * 2 * \left(\frac{\delta}{2.667} \right) * 4.619 \right\} * 3 + \left\{ M_p^{-ve} * \left(\frac{\delta}{2.309} \right) * 8.0 \right\} * 3 = 1843.68\delta$$

The external work will be the same as in the previous example. The collapse loads is then,

$$w_u^L = 199.62 \, \text{kN/m}^2$$

The *third example* is the same slab discussed before but subjected to a concentrated load placed at the center and it is required to obtain an upper-bound estimate of this load, P_u^L. As a result of this applied load, the slab will have both radial and tangential moments, which are expected to cause tension in the bottom in the neighborhood of the load and to some distance where the radial moment will reverse and cause tension in the top while the tangential moments will decay. Hence, the yield line pattern is expected to be as shown in Figure 3.41a. For simplicity, the bearing area of the load can be considered equal to zero, thus simplifying the mechanism to the shape in Figure 3.41b.

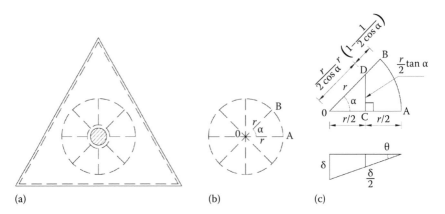

(a)　　　　　　　　　　　　(b)　　　　　　　　　　　　(c)

FIGURE 3.41 Example of a triangular slab under concentrated load.

Assume a vertical displacement δ under the load giving an external work $P_u^L \delta$. In order to calculate the dissipated energy, consider the segment OAB. For the negative plastic moment at AB, the angle of rotation is $\theta = \delta/r$ and the corresponding dissipated energy associated with this moment is

$$M_p^{-ve} * \left(\frac{\delta}{r} \right) * (r\alpha) * \frac{2\pi}{\alpha} = 2\pi\delta M_p^{-ve}$$

For the positive moment the rotation at the line OA is calculated at the middle of the line as follows (Figure 3.41c). The virtual displacement of point C is $\delta/2$ and the virtual displacement of point D is $\delta(1 - (1/2 \cos\alpha))$. The relative displacement between C and D is $(\delta/2)((1 - \cos\alpha)/\cos\alpha)$, and, therefore, the rotation of line CD is $(\delta/r)((1 - \cos\alpha)/\sin\alpha) = (\delta/r) \tan(\alpha/2) \approx (\delta\alpha/2r)$. The dissipated energy associated with positive moment is therefore

$$M_p^{+ve} * 2 * \left(\frac{\delta\alpha}{2r} \right) * r * \frac{2\pi}{\alpha} = 2\pi\delta M_p^{+ve}$$

The total dissipated energy is then $2\pi\delta(M_p^{+ve} + M_p^{-ve})$. Hence, the collapse load is

$$P_u^L = 2\pi(M_p^{+ve} + M_p^{-ve})$$

If the top reinforcement is zero, the same equation derived before applies with the substitution of $M_p^{-ve} = 0$.

It is worth repeating here that in the application of limit analysis to structural engineering, it is presupposed that, up to the instant of collapse, the deformation and the displacements remain sufficiently small so that one can ignore the change in the geometry of the deformed structure with the establishment and calculations of the equations of equilibrium on the element as well as its kinematical relationships. This restriction is not a problem for reinforced concrete structures since they are sufficiently rigid, but this may be an important limitation in the case of some thin-walled metal structures. However, for reinforced concrete material, it is necessary to consider the possibility of brittle fractures prior to collapse due to excessive yielding of steel. Strengthening the elements and avoiding brittle fractures require individual testing and verification before design.

It should also be mentioned that flexure is not the only failure mode but there are other modes as well such as failure by punching shear.

3.5   STRUT-AND-TIE MODEL AS A RECENT PROGRESS

3.5.1   INTRODUCTION

The upper-bound techniques of limit analysis, for example, the yield-line theory for slab design, as described in the preceding section, has long been used in engineering practice; however, the application of stress fields to reinforced concrete design based on the concept of lower-bound theorem of limit analysis is of more recent development. One of the most important advances in reinforced concrete in recent years is the extension of lower-bound-limit-theorem-based design procedures to shear, torsion, bearing stresses, and the design of structural discontinuities such as joints and corners.

The STM is based on the lower-bound theorem of limit analysis. In this model, the complex stress distribution in the structure is idealized as a truss carrying the imposed loading through the structure to its supports. Similar to a real truss, an STM consists of compression struts and tension ties interconnected at nodes. Using stress legs similar to those sketched in Figure 1.3, a lower-bound stress field that satisfies equilibrium and does not violate yield criteria at any point can be constructed to provide a safe estimate of capacity of reinforced concrete structures with discontinuities. As will be illustrated in the following examples, these techniques will have the advantage of allowing a designer to follow the forces through a structure with discontinuities, which formerly were beyond the scope of engineering practice.

The *STM* has been well developed in the United States over the last two decades and the subject was presented in several texts (Schlaich and Schäfer, 1991) as a standard method for shear, joints, and support bearing design. The *STM* method was also introduced in the *AASHTO LRFD Specifications* (1998) as well as in the *ACI 318 building codes* (2002, 2008). Chapter 5 attempts to make a simple, concise, and reasonably comprehensive introduction of this new theory for analysis and design of structural discontinuities in reinforced concrete structures.

3.5.2 CONCEPT

The STM is an idealization of the stress resultants derived from the flow of forces within a region of structural concrete. The successful model should satisfy two conditions: equilibrium and failure criteria. The solution so obtained is a safe or lower-bound solution.

The STMs are derived from the flow of forces within structural concrete regions, namely, those of high shear stresses, where Bernoulli hypothesis of flexure (plane sections before bending remain plane after bending) does not apply. Those regions are referred to as discontinuity or disturbance regions (or simply D-regions), in contrast to those regions where Bernoulli hypothesis is valid, and are referred to as Bernoulli or bending regions (or simply B-regions). The flow of forces in B-regions can be traced, of course, but in this case the model will yield to as the special case of the STM.

Discontinuity (which is associated with high shear stresses) is either static (as a result of concentrated loads) or geometric (as a result of abrupt change of geometry) or both. Examples of D-regions are illustrated in Figure 3.42. The dividing sections between B- and D-regions can be assumed to lie approximately at a distance *h* from the geometric discontinuity or the concentrated load, where *h* is equal to the thickness of the adjacent B-region (Figure 3.42). This assumption is justified by St. Venant's principle.

In an STM, a strut represents a concrete stress field with prevailing compression in the direction of the strut. On the other hand, a tie represents one or several

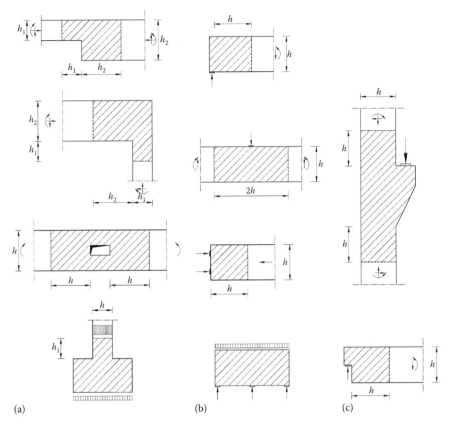

FIGURE 3.42 D-regions (shaded areas) with nonlinear strain distribution due to (a) geometric discontinuity, (b) static discontinuity, and (c) geometric and static discontinuities. (Adapted from Schlaich, J. et al., *J. Prestressed Concr. Inst.*, 32(3), 74, 1987; Schlaich, J. and Schäfer, K., *J. Struct. Eng.*, 69(6), 113, 1991; Schlaich, J. and Schäfer K., The design of structural concrete, *IABSE Workshop*, New Delhi, India, 1993.)

layers of tension reinforcement. However, concrete ties may exist in models where no reinforcement is available and reliance is on the concrete tensile strength. Examples where tensile stress fields are necessary for equilibrium can be traced in members such as slabs where no web reinforcement is used or in bar anchorage with no transverse reinforcement. Meanwhile, compression reinforcement is represented by a strut in case the need arises.

3.5.3 STRUT-AND-TIE MODELING

Before modeling a D-region the boundary forces acting from attached B-region or supports or external forces should be determined (Figure 3.43a). The stress diagrams of all forces applied to the D-region boundaries are subdivided in such a way that the individual stress resultants on opposite sides of the D-region correspond in magnitude and can be connected by streamlines that do not cross each other

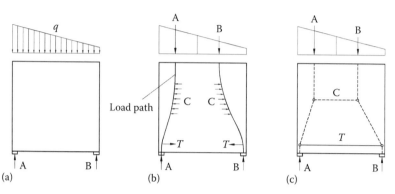

FIGURE 3.43   The load-path method (Schlaich and Schäfer, 1991): (a) the region and boundary loads, (b) the load paths through the region, and (c) the corresponding STM.

(Figure 3.43b). Then, the flow of forces through the region can be traced using the *load-path* method, Figure 3.43b, which are smoothly curved. Next, the load paths are replaced by polygons as shown in Figure 3.43c, and additional struts or ties are added for equilibrium, such as the transverse strut and tie in the figure. In some cases the stress diagrams or forces are not completely balanced with forces on the opposite side; for this, the load path of the remaining forces enters the structure and leaves it on the same side after a U turn within the region (Figure 3.44).

The development of an STM can be simplified if an elastic finite element (FE) analysis is performed to obtain the elastic stresses and principal stress directions (Schlaich and Schäfer, 1991). The location and direction of struts and ties can then be located at the center of stress diagrams (Figure 3.45). The orientation of struts and ties based on results from the theory of elasticity may not be the best choice in some cases where the profile and distribution of stresses may be altered as the load increases from working load level to collapse load with the associated nonlinear behavior of structural concrete. However, ductility of structural concrete may account for such a deviation. Also, the ties and hence the reinforcement may be arranged according to practical considerations; that is, the structure adapts itself to

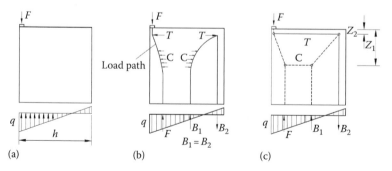

FIGURE 3.44   The load-path method including a U turn (Schlaich and Schäfer, 1991): (a) the region and boundary loads, (b) the load paths through the region, and (c) the corresponding STM.

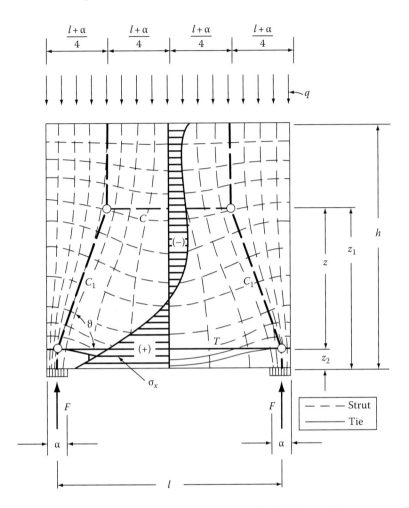

FIGURE 3.45 Elastic stress trajectories, elastic stress distribution, and the corresponding STM.

the assumed internal structural system. Nevertheless, modeling requires good design experience in order to set up proper design objectives such as safety and economy, and come up with a design that fulfills such objectives.

3.5.4 ELEMENTS OF STRUT-AND-TIE MODEL

An STM consists of three types of elements: struts, ties, and the connecting nodes or nodal zones (Figure 3.46). In the following, these elements are described in more details.

Strut: A strut is a compression member in an STM, which represents the resultant of a parallel or a fan-shaped compression field. In design, struts are usually idealized as prismatic compression members, as shown by the straight line outlines of the struts

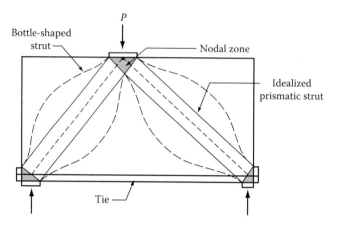

FIGURE 3.46 Description of STM.

in Figure 3.46. If the effective compression strength (or failure criterion) f_{cu} differs at the two ends of a strut, due either to different nodal zone strengths at the two ends, or to different bearing lengths, the strut is idealized as a uniformly tapered compression member.

Bottle-shaped strut: It is a strut that is wider at mid-length than at its ends, and it is located in a part of a member where the width of the compressed concrete at mid-length of the strut can spread laterally. The curved dashed outlines of the struts in Figure 3.46 and the curved solid outlines in Figure 3.47 approximate the boundaries of bottle-shaped struts. The internal lateral spread of the applied compression force in this stress field is similar to that of a split cylinder test. To simplify design, bottle-shaped struts are idealized either as prismatic or as uniformly tapered, and crack control reinforcement is provided to resist the transverse tension. The amount of

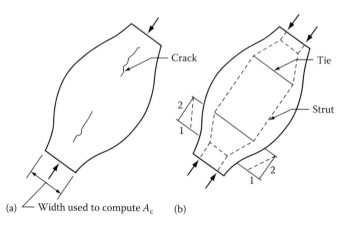

FIGURE 3.47 Bottle-shaped strut: (a) cracking of strut and (b) STM for transverse reinforcement.

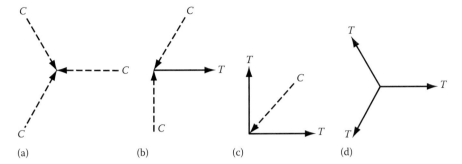

FIGURE 3.48 Nodes classification: (a) *C-C-C* node; (b) *C-C-T* node; (c) *C-T-T* node; and (d) *T-T-T* node.

confining transverse reinforcement can be computed using the STM shown in Figure 3.47, with the struts that represent the spread of the compression force acting at a slope of 1:2 to the axis of the applied compressive force. The cross-sectional area A_c of a bottle-shaped strut is taken as the smaller of the cross-sectional areas at the two ends of the strut. See Figure 3.47a.

Tie: It is a tension member in an STM where the force is resisted by normal reinforcement, prestressing, or concrete tensile strength. Within the scope of this discussion, only ties that represent normal reinforcement are considered. The reinforcement may consist of one or more layers and the force is always at the center of these layers.

Node: It is the point in a joint in an STM where the axes of the struts, ties, and concentrated forces acting on the joint intersect. For equilibrium, at least three forces should act on a node, as shown in Figure 3.48. Nodes are classified according to the signs of these forces. A *C-C-C* node resists three compressive forces; a *C-C-T* node resists two compressive forces and one tensile force, and so on.

Nodal zone: The volume of concrete around a node that is assumed to transfer strut-and-tie forces through the node is the nodal zone. Different types of nodal zones are illustrated in Figure 3.49 (ACI 318-08).

3.5.5 Failure Criteria

The different elements of an STM have to be checked or dimensioned according to the material failure criterion of the element. In literature there have been different assessments and different approaches for calculating strength values for elements of STMs. However, there is a noticeable variation and inconsistency between reported values and therefore the failure criteria adopted by the ACI 318-08 is adopted in this presentation.

3.5.5.1 Struts

The failure criterion of concrete in struts is denoted as f_{cu}, which is defined next. Therefore, the nominal strength of a strut is $f_{cu}A_c$, where A_c is the cross-sectional area at strut end. Hence, the smaller value of $f_{cu}A_c$ at the two ends of strut will control the design. In calculating A_c, the strut width w_s is measured perpendicular to the strut

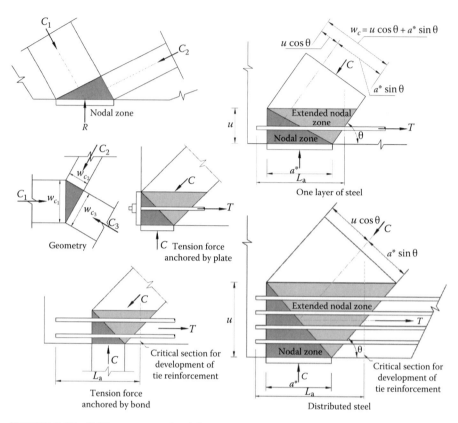

FIGURE 3.49   Different types of nodal zones.

axis at its end (Figure 3.49). f_{cu} shall not exceed the failure criterion of the node at the strut end into consideration:

$$f_{cu} = 0.85\beta_s f_c'$$ (3.89)

where

β_s is an effectiveness factor, which accounts for the shape of stress field and the associated conditions as will be illustrated in the following

f_c' is the concrete cylinder strength

$\beta_s = 1.00$ for strut of uniform cross-sectional area over its length

$\beta_s = 0.75$ for bottle-shaped strut when providing transverse reinforcement to resist the lateral tension according to the model in Figure 3.47, or if $f_c' \le 44\,\text{MPa}$ and the reinforcement crossing the strut, Figure 3.50, satisfy $\Sigma(A_{s_i}/bs_i)\sin\gamma_i \ge 0.003$, where A_{s_i} is the total area of reinforcement at spacing s_i in a layer of reinforcement with bars at an angle γ_i to the axis of the strut and the other parameters are as illustrated in the figure

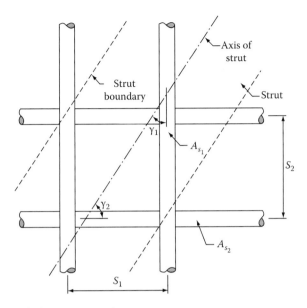

FIGURE 3.50   Reinforcement crossing a strut.

$\beta_s = 0.60$ for bottle-shaped strut when the transverse reinforcement does not satisfy the requirement of the model in Figure 3.47. This value is assigned to normal strength concrete

$\beta_s = 0.40$ for struts in tension members or the tension flanges of members

$\beta_s = 0.60$ for all other cases, for example, struts in a beam web compression field in the web of a beam where parallel diagonal cracks are likely to divide the web into struts, and struts are likely to be crossed by cracks at an angle to the struts (Figure 3.51)

3.5.5.2   Ties

The failure criterion of ties representing normal reinforcement is the steel yield stress, σ_o. Therefore, the nominal strength of a tie is $\sigma_o A_s$, where A_s is the cross-sectional area of the reinforcing steel.

3.5.5.3   Nodal Zones

The failure criterion of concrete in nodal zones is denoted as f_{cu}, which is defined next. Hence, the nominal strength of a nodal zone is $f_{cu} A_n$, where A_n is the area of the face of the nodal zone that F_u acts on, taken perpendicular to the line of action of F_u, where F_u is the factored force acting at the nodal zone section. Alternatively, A_n is the area of a cross section through the nodal zone, taken perpendicular to the line of action of the resultant force on that section:

$$f_{cu} = 0.85\beta_n f_c' \qquad (3.90)$$

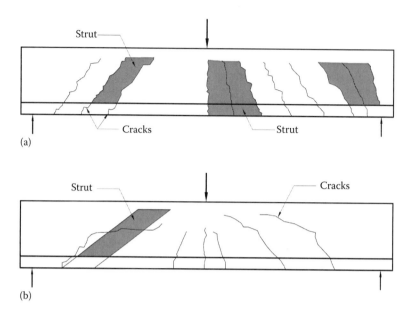

FIGURE 3.51 Types of struts in a beam web: (a) struts in a beam web with inclined cracks parallel to struts and (b) struts crossed by skew cracks.

where β_n is a factor, which reflects the increasing degree of disruption of the nodal zones due the incompatibility of tension strains in the struts.

$\beta_n = 1.00$ for nodal zones bounded by struts or bearing areas or both
$\beta_n = 0.80$ for nodal zones anchoring one tie
$\beta_n = 0.60$ for nodal zones anchoring two or more ties

3.5.6 AN ILLUSTRATIVE EXAMPLE

It is required to determine the reinforcement for the simply supported transfer girder shown in Figure 3.52. The single column at the mid-span carries a factored load 2800 kN. The concrete cylinder strength is $f'_c = 32$ MPa and the steel yield stress is $\sigma_o = 420$ MPa. Neglect the beam's own weight. The strength reduction factor for the effective concrete of struts and nodes and for ties yield stress is 0.75. The solution is given by the following steps.

Reactions

$$R_A = R_B = 1400\,\text{kN}$$

Establish an STM

In this beam, the shear span to depth ratio is less than 2; therefore, the beam is considered a D-region, that is, deep beam. The appropriate STM is shown in Figure 3.53, in which the lower nodes are assumed to coincide with the centerlines of the

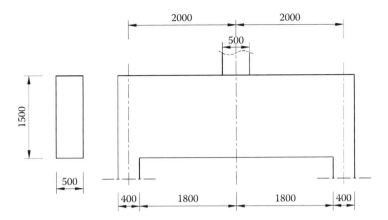

FIGURE 3.52 Transfer girder example.

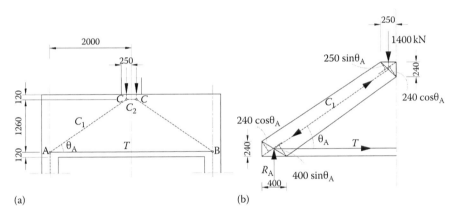

(a)   (b)

FIGURE 3.53 STM of the transfer girder.

supporting columns and all nodes are located 120 mm from the upper and lower edges of the beam as shown in the figure.

With reference to Figure 3.53a, the length of the diagonal strut is given by

$$\sqrt{(1260)^2 + (2000 - 125)^2} = 2259\,\text{mm}$$

The force in strut C_1 is $1400 \times \dfrac{2259}{1260} = 2510\,\text{kN}$

The force in tie T is $1400 \times \dfrac{2000 - 125}{1260} = 2083\,\text{kN}$

The force in strut C_2 is $2083\,\text{kN}$

The angle between C_1 and T, $\theta_A = \tan^{-1}\dfrac{1260}{2000 - 125} = 33.9° > 30.0°$, O.K.

Effective concrete strength for the struts

For strut C_1, $f_{cu} = 0.85 \times 0.75^* \times f'_c = 20.40\,\text{MPa}$

For strut C_2, $f_{cu} = 0.85 \times 1.0 \times f'_c = 27.20\,\text{MPa}$

Effective concrete strength for the nodes

For Node A, $f_{cu} = 0.85 \times 0.80 \times f'_c = 21.76\,\text{MPa}$
For Node C, $f_{cu} = 0.85 \times 1.0 \times f'_c = 27.20\,\text{MPa}$

Node C

The bearing stress $= \dfrac{1400 \times 10^3}{250 \times 500} = 11.20\,\text{MPa} < 0.75^\dagger \times 27.20\,\text{MPa}\,(= 20.40\,\text{MPa})$

The required width of strut C_2, $w_{C_2} = \dfrac{2083 \times 10^3}{500 \times (0.75 \times 27.20)} = 204\,\text{mm}$

The difference between the assumed width (240 mm) and the required width (204 mm) is on the safe side and it is not significant; therefore, the solution will proceed without modifying the dimensions of the STM. Since node C is under hydrostatic pressure and two sides are safe, the third side of the node (the side of strut C_1) will also be safe. However, for illustration it is checked here. With reference to Figure 3.53b, the width of strut C_1 is

$$w_{C_1} = 250 \sin \theta_A + 240 \cos \theta_A = 339\,\text{mm}$$

which corresponds to a stress equal to

$$\frac{2510 \times 10^3}{339 \times 500} = 14.81\,\text{MPa} < 0.75 \times 27.20\,\text{MPa}\ (= 20.40\,\text{MPa})$$

Node A

The bearing stress is equal to

$$\frac{1400 \times 10^3}{400 \times 500} = 7.00\,\text{MPa} < 0.75 \times 21.76\,\text{MPa}\ (= 16.32\,\text{MPa})$$

The width of strut C_1 is

$$w_{C_1} = 240 \cos \theta_A + 400 \sin \theta_A = 422\,\text{mm}$$

which corresponds to a stress equal to

$$\frac{2510 \times 10^3}{422 \times 500} = 11.90\,\text{MPa} < 0.75 \times 21.76\,\text{MPa}\ (= 16.32\,\text{MPa})$$

---

* Transverse reinforcement to resist the lateral tension will be provided.
† The strength reduction factor.

Strut C_1

The width of the strut at the end A is 422 mm and at the end C is 339 mm; the smaller value is considered in checking the safety of the strut. The stress in the strut is therefore equal to

$$\frac{2510 \times 10^3}{500 \times 339} = 14.82\,\text{MPa} < 0.75 \times 20.40\,\text{MPa} (= 15.30\,\text{MPa})$$

The transverse reinforcement of the strut is required to resist a total force T_{C_1}. From the STM of Figure 3.47,

$$T_{C_1} = \left(\frac{1}{2} \times \frac{C_1}{2}\right) \times 2 = \frac{C_1}{2} = \frac{2510}{2} = 1255\,\text{kN}$$

Thus, the total required reinforcement in perpendicular to the strut is $1255 \times 10^3/0.75 \times 420 = 3984\,\text{mm}^2$. The length of the strut is 2259 mm; hence, the required transverse reinforcement is $1.764\,\text{mm}^2/\text{mm}$, in perpendicular to the strut. This can be covered with a skin reinforcement of vertical bars of diameter 16 mm every 200 mm and horizontal bars of diameter 12 mm every 200 mm, on each side, in addition to interior open stirrups of diameter 10 mm every 400 mm. The larger diameter is assigned to the vertical bars since they are more effective in substituting for the inclined reinforcement because the angle θ_A is less than 45°. With reference to Figure 3.50, the used transverse steel is equivalent to inclined reinforcement $\Sigma A_{si} \sin \gamma_i / bs_i = 0.0046 > 0.003$ (the ACI minimum value for $f_c' \leq 44\,\text{MPa}$).

Strut C_2

Since the effective concrete strength of this strut is the same as the end node and the node is safe, the strut is safe.

Tie T

The reinforcement required to resist the force of this tie is $2083 \times 10^3/0.75 \times 420 = 6613\,\text{mm}^2$, which can be covered with 14 bar of diameter 25 mm. This reinforcement should be extended in the node and beyond the anchorage length required to develop the force in the tie.

3.6 HISTORICAL SKETCH

The theory of plasticity began when the experiments of Tresca (1870) in the 1860s established the concept that a large plastic deformation is shear deformation governed primarily by shear stress. It took great insight and a giant step forward by St. Venant (1870) and Levy (1870) to propose their plastic stress–strain relations by ignoring the relatively small elastic strain increments and considering only the plastic strain increments. Also, the initial state of the material was ignored in this simplest perfectly plastic idealization. Much later, Prandtl (1924) and Reuss (1930) added an isotropic elastic response to the St. Venant–Levy equations to produce the simplest incremental stress–incremental strain relations for an isotropic elastic–plastic material.

Meanwhile, and subsequently, many experimenters and those who analyzed their results attempted to refine and understand the yield criterion or the commonly called failure criterion at the time. These new refined theories or criteria include, for example, maximum strain energy theory, maximum shear strain energy theory, and maximum tensile strain theory, among others. von Mises (1913) accepted the Tresca criterion and proposed a simple mathematical function called octahedral shear stress, or shear strain energy, or J_2, to represent the von Mises criterion for mathematical convenience. Nowadays, the von Mises yield criterion also bears the names of Huber and Hencky. Subsequently, von Mises (1928) proposed a general normality rule for the plastic increment of strain with any choice of smooth isotropic or anisotropic function of stress called potential function, which could be but was not necessarily identical to or even similar to the yield function of the material.

Once the beginning of a mathematically attractive and physically admissible base had been established for the idealization of stress–strain relations in the plastic range, it is natural to extend these relations to include the major physical feature of work-hardening of the material as a logical next step. A J_2 or von Mises isotropic hardening rule proposed by Odqvist (1933) did just that. The subsequent generalization of isotropic stress hardening by Melan (1938) to include the third invariant of the stress deviation tensor, J_3, represented a further step forward in continuum mechanics.

The modern approach to the mathematical theory of plasticity was best described by the two survey papers by Prager (1948, 1955). The surveys introduced many concepts now familiar in plasticity theory, most notably to structural engineers in particular, including the concept of generalized stresses and generalized strains, the principle of maximum plastic work, the equivalent convexity for yield function and normality condition for plastic strain rates at yield, the description of kinematic hardening for the yield surface in stress space to exhibit Bauschinger and allied effects, and the proof of the theorems of limit analysis, among others. The great outpouring of these theoretical works at Brown University under the leadership of Prager in the 1950s, and the subsequent development of plastic design methods for steel frames at Cambridge University (J. F. Baker) and Lehigh University (L. S. Beedle) in the 1960s started the revolutions. Many useful results, many ways of thinking about machines, structures, and materials as continua have come out of the plastic analysis and limit theorems.

Structural applications of plastic analysis and limit design in steel (Chen and Atsuta, 2007, Chen and Toma, 1994, Chen and Sohal, 1995; Chen et al., 1996), in soils (Chen, 1975; Chen and Baladi, 1985; Chen and Liu, 1990; Chen and Mizuno, 1990) and in concrete (Chen, 1982, 1994) are complemented by applications to metal forming processing (Johnson, 1986) in recent decades. Chen has been particularly active in the development of constitutive equations for engineering materials suitable for FE types of applications (Chen, 1994); especially the publication of *Plasticity for Structural Engineers* by Chen and Han in 1988 made the highly mathematical theory of plasticity easy for structural engineers to understand and master. These equations are useful for determining the history of the stress and the strain in a structure when the history of loading and constraint is known for the problem.

In the following, we give a very abbreviated version of a long story of great leaps forward of the development of the modern theory of plasticity by focusing on the introduction of the life and career of two pioneering giants in plasticity at Brown University, William Prager and his close associate Daniel C. Drucker to whom the senior author, W. F. Chen, had the great fortune to be his graduate student during the period 1963–1966 and later to be associated with him professionally until his death in 2001.

The Brown group in solid mechanics was founded by William Prager, and through his leadership, the group became internationally known for its pioneering work in plasticity. It was at Brown University that Prager and Drucker made the major part of their contributions to the theory of plasticity and to its application to the design of engineering structures and components. They were the coauthors of the current classical paper in which the limit load was clearly defined and the theorems of limit analysis were established (Drucker et al., 1952). The theorems led directly to limit design—a technique to predict the load-carrying capacity of engineering structures and components such as bridges, pressure vessels, and machine parts. The theorems had immediate applications to problems, which formerly were beyond the scope of engineering practice.

William Prager (O'Connor and Robertson, 2005) was born in Karlsruhe in Germany in 1903 and obtained his Dipl. Ing. degree in 1925 and doctorate in engineering in 1926 from the Technical University of Darmstadt. He remained at Darmstadt as an instructor and in 1929 he was appointed to act as director of the Institute of Applied Mathematics at Göttingen. As a result of his leading international reputation, with over 30 papers and a book already showing the depth of his contributions to applied mathematics, Prager was appointed as professor of technical mechanics at Karlsruhe in 1932, to become the youngest professor in Germany.

In 1934, Prager left Germany for Turkey because of the Nazi regime, where he was appointed as professor of theoretical mechanics at the University of Istanbul. In Turkey he continued to produce research at the highest level, publishing articles in German, Turkish, French, and English. The outbreak of war in 1939 was distressing to Prager and the German advances by 1940 made him decide that he would be best placed if he could emigrate to the United States. In 1941, Brown University took the opportunity to expand its graduate program by offering Prager the position of director of Advanced Instruction and Research in Mechanics.

Prager established the Division of Applied Mathematics at Brown in 1946, served as its first chairman, and guided its research and teaching by gathering around him younger people in a wide variety of fields of applied mechanics, applied mathematics, physics, and engineering. His research during this period covered an enormous diversity of topics in the mechanics of continua of all types, problems of traffic flow, and applications of computers to problems in economics and engineering. J. L. Synge was a visiting professor at Brown University in 1941 when Prager arrived there. They soon began collaborating and publishing papers in the *Quarterly of Applied Mathematics*, which Prager founded in April 1943 and edited for over 20 years. In the Walker-Ames Lectures, Prager developed the hypercircle method, applying it to statically indeterminate structures and to the equilibrium of elastic solids; the lectures were published as the extremum principles of the mathematical theory of elasticity and their use in stress analysis in 1950. An important monograph at that time, which Prager wrote jointly with P. G. Hodge, was *Theory of Perfectly Plastic Solids* (1951).

In November and December 1954 Prager gave a series of lectures at the Polytechnic Institute in Zurich. These were published in the following year as *Probleme der Plastizitätstheorie*. E. T. Onat reviewing the book writes:

> The book constitutes a clear and penetrating exposition of the concepts and applications of the theory of plasticity. The author is one of the principal contributors in the field and his book provides the reader with indications of the impending developments of the theory.

Prager further developed the material given in these lectures and presented it in an English version in an *Introduction to Plasticity* published in 1959. A review of this book, this time by J. Heyman, again gives Prager high praise:

> There are no spare lines and there is no padding; the author has considered every word, and thought deeply on every aspect of plastic theory. ... the author is completely master of this, his main field of study, and he communicates this sense of mastery to the reader.

In 1961 Prager published a German and an English version of the same work.

Prager retired from Brown University in 1973 and moved to Savignon, Switzerland, where he continued to undertake research, write books, give lecture tours, and edit journals. In particular, he gave six lectures at the International Centre for Mechanical Sciences in Udine in 1974, which he wrote up and published as *Introduction to Structural Optimization* (1974). His first three lectures considered the derivation of necessary and sufficient conditions for global optimality from extremum principles, while the final three lectures looked at the optimization of the structural layout.

His outstanding contributions to applied mathematics led to Prager receiving many honors and awards. He was elected to the National Academy of Engineering, the National Academy of Sciences, the American Academy of Arts and Sciences, the Polish Academy of Sciences, and the French Académie des Sciences. He received the Worcester Reed Warner medal and the Timoshenko medal from the American Society of Mechanical Engineers and the von Karman medal from the American Society of Civil Engineers. Many universities awarded him honorary degrees including Liege, Poitiers, Milan, Waterloo, Stuttgart, Hannover, Brown, Manchester, and Brussels.

Dan Drucker was born in New York City in 1918, attended Columbia University, where he earned three degrees, including a PhD at the age of 21. His doctoral thesis on three-dimensional photo-elastic methods was under the supervision of Professor R. D. Mindlin. He went on to teach at Cornell University and worked at the Armour Research Foundation before spending a year in the U.S. Army Air Corps during World War II.

In 1947, Drucker went to Brown University where for two decades he helped build one of the best programs in the country in materials engineering and solid mechanics. He is best known for his pioneering work in the theory of plasticity and its applications to analysis and design of metal structures. For almost half a century, the advances in the theory of plasticity and the work of Drucker were intimately linked. The familiar terms such as stability postulate, stable materials, limit theorems, limit analysis, plastic design, Drucker–Prager model, and soil plasticity come to mind immediately, among others. He introduced the concept of material stability, now known as "Drucker's Stability Postulate," which provided a unified approach for the derivation of stress–strain relations for the plastic behavior of metals. The simple fact that his name is attached to it and that it has survived for half century is indicative of the significance of the person.

Drucker was the first to show how limit analysis could be used in the design of cylindrical shells, and later he applied it effectively to the design of pressure vessels. His plasticity work also extended to include soil mechanics, metal working, and metal cutting. In his later years at Brown, he became active in the field sometimes known as micro-mechanics, which attempts to bridge the gap between the material scientists who study material behavior at the atomic level and the engineers who work with real materials modeled by theories of continuum mechanics.

In 1968, Drucker became the dean of engineering at the University of Illinois, where he is credited with improving the quality of the faculty. In 1984, Drucker left Illinois to become a graduate research professor at the University of Florida, where he was a kind of senior statesman on campus and was always available to help, especially the younger faculty members.

Among his life achievements, Drucker was honored with almost every major award given by major engineering societies including the von Karman medal from ASCE, Timoshenko medal from ASME, and Lamme medal from ASEE. He was a member of the National Academy of Engineering, received honorary doctorates from five different universities, and served as president of five U.S. and international societies. One of Drucker's most prestigious honors was given by a U.S. President—the National Medal of Science.

REFERENCES

AASHTO, 1998, *AASHTO LRFD Bridge Specifications*, 2nd edn., American Association of State Highway and Transportation Officials, Washington, DC.

ACI 318-02, 2002, *Building Code Requirements for Structural Concrete and Commentary*, American Concrete Institute, Farmington Hills, MI.

ACI 318-08, 2008, *Building Code Requirements for Structural Concrete and Commentary*, American Concrete Institute, Farmington Hills, MI.

Chen, W. F., 1975, *Limit Analysis and Soil Plasticity*, Elsevier, Amsterdam, the Netherlands, Reprinted by J. Ross, Orlando, FL, 2007.

Chen, W. F., 1982, *Plasticity in Reinforced Concrete*, McGraw-Hill, New York.

Chen, W. F., 1994, *Constitutive Equations for Engineering Materials*, Elsevier, Amsterdam, the Netherlands.

Chen, W. F. and Atsuta, T., 2007, *Theory of Beam-Columns, Vol. 1—In-Plane Behavior and Design*, McGraw-Hill, New York, Reprinted by J. Ross, Orlando, FL, 1976.

Chen, W. F. and Atsuta, T., 2007, *Theory of Beam-Columns, Vol. 2—Space Behavior and Design*, McGraw-Hill, New York, Reprinted by J. Ross, Orlando, FL, 1977.

Chen, W. F. and Baladi, G. Y., 1985, *Soil Plasticity: Theory and Implementation*, Elsevier, Amsterdam, the Netherlands.

Chen, W. F., Goto, Y., and Liew, J. Y. R., 1996, *Stability Design of Semi-Rigid Frames*, John Wiley & Sons, New York.

Chen, W. F. and Han, D. J., 1988, *Plasticity for Structural Engineers*, Springer-Verlag, New York.

Chen, W. F. and Kim, S. E., 1997, *LRFD Design Using Advanced Analysis*, CRC Press, Boca Raton, FL.

Chen, W. F. and Liu, X. L., 1990, *Limit Analysis in Soil Mechanics*, Elsevier, Amsterdam, the Netherlands.

Chen, W. F. and Mizuno, E., 1990, *Nonlinear Analysis in Soil Mechanics*, Elsevier, Amsterdam, the Netherlands.

Chen, W. F. and Sohal, I., 1995, *Plastic Design and Second-Order Analysis of Steel Frames*, Springer-Verlag, New York.

Chen, W. F. and Toma, S., 1994, *Advanced Analysis for Steel Frames: Theory, Software and Applications*, CRC Press, Boca Raton, FL.

Coulomb, C. A., 1773, Essai sur une application des regles de maximis et minimis a quelque problems de statique relatifs a l'architecture, *Memoires Math. Acad. Royale Sci. Paris*, 7, 343–382.

Drucker, D. C., Prager, W., and Greenberg, H., 1952, Extended limit design theorems for continuous media, *Quarterly Applied Mathematics*, 9, 381–389.

ECCS, 1991, *Essential of Eurocode 3 Design Manual for Steel Structures in Buildings*, ECCS-Advisory Committee 5, No. 65.

Jirasek, M. and Bazant, Z. P., 2002, *Inelastic Analysis of Structures*, John Wiley & Sons, Ltd, New York.

Johnson, W., 1986, Mathematical programming applications in engineering plastic analysis, in: *Applied Mechanics Update 1986* (C. R. Steele and G. S. Springer, eds.), pp. 157–175.

Levy, M., 1870, Memoire sur les equations generales des mouvements interieurs des corps solides ductiles au dela des limites ou l'elasticite pourrait les ramener a leur premier etat, *C. R. Acad. Sci. Ser. II*, 70, 1323–1325.

Liew, J. Y. R., White, D. W., and Chen, W. F., 1994, Notional-load plastic-hinge method for frame design, *Journal of Structural Engineering, ASCE*, 120, 5, 1434–1454.

MacGregor, J. G. and Wight, J. K., 2005, *Reinforced Concrete Mechanics and Design*, Pearson, Prentice Hall, NJ.

Melan, E., 1938, Zur Plastizitat des raumlichen kontinuums, *Ingenieur-Arch*, 9, 116–126.

Mises, R. v., 1913, Mechanik der festen Korper im plastisch deformablen Zustand, Nachrichten der Gesellschaft der Wissenshaften Gottingen, *Math-phys Klasse*, 582–592.

Mises, R. v., 1928, Mechanik der plastischen Formanderung von Kristallen, *ZAMM*, 8, 161–185.

O'Connor, J. J. and Robertson, E. F., 2005, William Prager, www-groups.dcs.st-and.ac.uk/~history/Biographies/ accessed on 11/19/2010.

Odqvist, F. K. G., 1933, Die Verfestigung von flusseisenahnlichen Korpern, Ein Beitrag zur Plastizitatstheorie, *ZAMM*, 13, 360–363.

Prager, W., 1948, The stress-strain laws of the mathematical theory of plasticity: A survey of recent progress, *Journal of Applied Mechanics*, 15, 226–233.

Prager, W., 1955, The theory of plasticity—A survey of recent achievements, *Proceedings of the Institution of Mechanical Engineers*, 169, 41–57.

Prager, W., 1955, Probleme der Plastizitatstheorie. Available at: http://www-history.mcs.st-and.ac.uk/Biographies/Prager.html

Prager, W., 1959, *Introduction to Plasticity*, Addison-Wesley, Reading Mass.

Prager, W., 1974, *Introduction to Structural Optimization*, CISM 212, Springer, Vienna.

Prager, W. and Hodge, P. G., Jr., 1951, *Theory of Perfectly Plastic Solids*, Wiley, New York.

Prandtl, L., 1924, Spannungsverteilung in plastischen Korpern, *Proceedings of the First International Congress of Applied Mechanics*, Delft, 1924, edited by C.B. Biezeno and J.M. Burgers (one Vol).

Reuss, A., 1930, Berucksichtung der elastischen Formanderung in der Plastizitatstheorie, *ZAMM*, 10, 266–274.

Schlaich, J. and Schäfer, K., 1991, Design and detailing of structural concrete using strut- and-tie models, *Journal of the Structural Engineer*, 69, 6.

Schlaich, J. and Schäfer, K., 1993, The design of structural concrete, *IABSE Workshop*, New Delhi, India.

Schlaich, J., Schäfer, K., and Jennewein, M., 1987, Toward a consistent design of structural concrete, *Journal of the Prestressed Concrete Institute*, 32, 3, 74–150.

Sokolovsky, V. V., 1946, *Theory of Plasticity*, Moscow, Russia.

St. Venant, B. de, 1870, Memoire sur l'establissement des equations differentialles des mouvements interieurs operes dans les corps solides ductiles au dela des limites ou l'elasticite pourrait les ramener a leur premier etat, *C. R. Acad. Sci. Ser. II*, 70, 473–480.

Tresca, H., 1870, Memoire sur le poinconnage des metaux et des matieres plastiques, *C. R. Acad. Sci. Ser. II*, 70, 27–31.

Wong, M. B., 2009, *Plastic Analysis and Design of Steel Structures*, Elsevier, Amsterdam, the Netherlands.

4 The Era of Finite Element

4.1 INTRODUCTION

Modern computational techniques, and, in particular, the finite element method (FEM) have been well developed and used widely in nonlinear analysis of structures since the 1970s. Thanks to this success, we were able to apply the theory of stability and the theory of plasticity to simulate the actual behavior of structural members and frames with great confidence. It was the first time we were able to replace the costly full-scale tests with computer simulation. As a result, the limit state approach to design was advanced and new specifications were issued. The state of the art in finite element (FE) modeling of structural elements with properties of materials and kinematic assumptions is examined in this chapter along with a brief description of the impacts of the applications of this method on structural engineering practice.

4.2 FUNDAMENTALS OF FINITE ELEMENT

4.2.1 KINEMATICS CONDITIONS (SHAPE FUNCTION)

In the FEM, formulation starts with the kinematical (compatibility) conditions. In this step, the generic displacements of an element (internal displacements within an element), $\{u\}$, are related to the generalized strains (the nodal displacement of the element), $\{q\}$, by means of assumed shape function, $[N]$. This assumption is equivalent to Bernoulli's assumption in beams under bending (plane sections perpendicular to the neutral axis before bending remain plane and perpendicular to the neutral axis after bending) or Kirchhoff's hypothesis in plates under bending:

$$\{u\} = [N]\{q\} \tag{4.1}$$

With the displacement within the element, the strain vector, $\{\varepsilon\}$, at any point within the element can be obtained by differentiation of $\{u\}$ in Equation 4.1

$$\{\varepsilon\} = [B]\{q\} \tag{4.2}$$

where $[B]$ is composed of derivatives of the shape function, $[N]$.

The development of matrices $[N]$ and $[B]$ is illustrated here for the constant strain triangle element. Consider the cantilever plate in Figure 4.1a subjected to an in-plane load, and, therefore, it is in a state of plane stress. The plate can be idealized as an assemblage of two-dimensional plane stress FEs, as shown in Figure 4.1b. In order to develop the matrix $[N]$ of an element, it is assumed that the element-generic displacements (referred to sometimes as element-local displacements), u and v, are given in the form of polynomials in the local coordinate variables x and y (Yang, 1986):

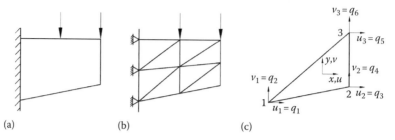

FIGURE 4.1 Plane stress problem: (a) structural element; (b) discretization; and (c) finite element.

$$u(x, y) = c_1 + c_2 x + c_3 y \tag{4.3a}$$

$$v(x, y) = c_4 + c_5 x + c_6 y \tag{4.3b}$$

One of the reasons for assuming such displacement functions is that six constants $c_1,...c_6$ can be determined uniquely by using the six nodal displacements $\{q\}$, Figure 4.1c, where

$$\{q\}^t = \{u_1 \ v_1 \ u_2 \ v_2 \ u_3 \ v_3\} = \{q_1 \ q_2 \ q_3 \ q_4 \ q_5 \ q_6\} \tag{4.4}$$

From Equations 4.3 and 4.4, and with reference to Figure 4.1c,

$$\{q\} = \begin{Bmatrix} q_1 \\ q_2 \\ q_3 \\ q_4 \\ q_5 \\ q_6 \end{Bmatrix} = \begin{bmatrix} 1 & x_1 & y_1 & 0 & 0 & 0 \\ 0 & 0 & 0 & 1 & x_1 & y_1 \\ 1 & x_2 & y_2 & 0 & 0 & 0 \\ 0 & 0 & 0 & 1 & x_2 & y_2 \\ 1 & x_3 & y_3 & 0 & 0 & 0 \\ 0 & 0 & 0 & 1 & x_3 & y_3 \end{bmatrix} \begin{Bmatrix} c_1 \\ c_2 \\ c_3 \\ c_4 \\ c_5 \\ c_6 \end{Bmatrix} \tag{4.5}$$

or

$$\{q\} = [A]\{c\} \tag{4.6}$$

The constant matrix can then be derived:

$$\{c\} = [A]^{-1}\{q\} \tag{4.7}$$

The generic displacement will be

$$\{u\} = \begin{Bmatrix} u \\ v \end{Bmatrix} = \begin{bmatrix} 1 & x & y & 0 & 0 & 0 \\ 0 & 0 & 0 & 1 & x & y \end{bmatrix} [A]^{-1}\{q\} \tag{4.8}$$

or

$$\{u\} = \begin{Bmatrix} u \\ v \end{Bmatrix} = [N]\{q\} \tag{4.9}$$

where

$$[N] = \begin{bmatrix} 1 & x & y & 0 & 0 & 0 \\ 0 & 0 & 0 & 1 & x & y \end{bmatrix}[A]^{-1} \tag{4.10}$$

For the derivation of matrix $[B]$ recall the strain–displacement relation

$$\{\varepsilon\} = \begin{Bmatrix} \varepsilon_x \\ \varepsilon_y \\ \gamma_{xy} \end{Bmatrix} = \begin{bmatrix} \dfrac{\partial u}{\partial x} & 0 \\ 0 & \dfrac{\partial v}{\partial y} \\ \dfrac{\partial u}{\partial y} & \dfrac{\partial v}{\partial x} \end{bmatrix} = \begin{bmatrix} \dfrac{\partial}{\partial x} & 0 \\ 0 & \dfrac{\partial}{\partial y} \\ \dfrac{\partial}{\partial y} & \dfrac{\partial}{\partial x} \end{bmatrix} \begin{Bmatrix} u \\ v \end{Bmatrix} = [d]\{u\} \tag{4.11}$$

From Equations 4.2, 4.9, and 4.11,

$$[B] = [d][N] \tag{4.12}$$

where $[d]$ is the differential operator given by

$$[d] = \begin{bmatrix} \dfrac{\partial}{\partial x} & 0 \\ 0 & \dfrac{\partial}{\partial y} \\ \dfrac{\partial}{\partial y} & \dfrac{\partial}{\partial x} \end{bmatrix} \tag{4.13}$$

Thus,

$$[B] = [d][N] = \begin{bmatrix} \dfrac{\partial}{\partial x} & 0 \\ 0 & \dfrac{\partial}{\partial y} \\ \dfrac{\partial}{\partial y} & \dfrac{\partial}{\partial x} \end{bmatrix} \begin{bmatrix} 1 & x & y & 0 & 0 & 0 \\ 0 & 0 & 0 & 1 & x & y \end{bmatrix}[A]^{-1} = \begin{bmatrix} 0 & 1 & 0 & 0 & 0 & 0 \\ 0 & 0 & 0 & 0 & 0 & 1 \\ 0 & 0 & 1 & 0 & 1 & 0 \end{bmatrix}[A]^{-1} \tag{4.14}$$

It is noted that the elements of matrix $[B]$ are constant terms. With reference to Equation 4.2, this means that any strain component will be a constant value throughout the element.

In order to guarantee compatibility in small, the interpolation function of the generic displacement should always be assumed polynomial with a form that depends on the element type. For instance, for a bilinear strain rectangle element, the displacements u and v are assumed as

$$u(x, y) = c_1 + c_2 x + c_3 y + c_4 xy \tag{4.15a}$$

$$v(x, y) = c_5 + c_6 x + c_7 y + c_8 xy \tag{4.15b}$$

For plate bending, the displacement w of the ACM (Adini, Clough, and Melosh) element (Desai and Abel, 1972) is

$$w(x, y) = c_1 + c_2 x + c_3 y + c_4 x^2 + c_5 xy + c_6 y^2 + c_7 x^3 + c_8 x^2 y$$

$$+ c_9 xy^2 + c_{10} y^3 + c_{11} x^3 y + c_{12} xy^3 \tag{4.16}$$

The assumption of the interpolation function in the form of a polynomial guarantees that compatibility within the element (compatibility in small) is satisfied. However, inter-element compatibility (compatibility in large) may not be satisfied as in the ACM element.

4.2.2 Equilibrium Conditions (Principle of Virtual Work)

Equilibrium conditions are imposed in order to obtain the relation between the generalized stresses (nodal force vector), $\{F\}$, and the internal stress vector at any point, $\{\sigma\}$. This can be achieved upon the application of the principle of virtual work.

In the principle of virtual work, it is assumed that the virtual displacements are so small that there will be no significant change in geometry. Thus, the forces may also be assumed to remain unchanged during the virtual displacements. For an element subjected to a system of loads $\{P\}$ and undergoing virtual nodal displacements $\{\delta q\}$, the external work done δW^* is

$$\delta W^* = \{\delta q\}'\{P\} \tag{4.17}$$

The applied loads result in a stress vector $\{\sigma\}$ and the virtual displacements $\{\delta q\}$ are associated with virtual strains $\{\delta \varepsilon\}$ by the equation

$$\{\delta \varepsilon\} = [B]\{\delta q\} \tag{4.18}$$

The virtual strain energy density is $\{\delta \varepsilon\}'\{\sigma\}$ and hence the virtual strain energy δU^* is

$$\delta U^* = \int_V \{\delta\varepsilon\}^t \{\sigma\} \, dV = \int_V \{\delta q\}^t [B]^t \{\sigma\} \, dV = \{\delta q\}^t \int_V [B]^t \{\sigma\} \, dV \qquad (4.19)$$

According to the principle of virtual work, $\delta W^* = \delta U^*$; hence, from Equations 4.17 and 4.19,

$$\{P\} = \int_V [B]^t \{\sigma\} \, dV \qquad (4.20)$$

Equation 4.20 provides the relation between the element stresses and the element-generalized stresses (nodal loads).

In achieving equilibrium condition, the principle of virtual work has been utilized, which demonstrates the fact that the principle of virtual work is nothing other than a principle of equilibrium. In this regard, equilibrium is satisfied in large but not in small. In other words, the overall element equilibrium is satisfied but a tiny portion of the element may not be in equilibrium. It is also realized that equilibrium along the boundaries between elements may not be satisfied.

4.2.3 Constitutive Conditions (Incremental/Iterative Formulation)

The final step in the formulation of the FE equilibrium equations is to impose the constitutive conditions. This is illustrated next for linear elastic analysis followed by nonlinear analysis.

For *linear elastic analysis*, the constitutive relations can be expressed in the following general form:

$$\{\sigma\} = [C]\{\varepsilon\} \qquad (4.21)$$

where $[C]$ is called the elastic constitutive or elastic moduli matrix. Upon substituting (4.2) and (4.21) into (4.20),

$$\{P\} = [k]\{q\} \qquad (4.22)$$

where

$$[k] = \int [B]^t [C][B] \, dV \qquad (4.23)$$

The matrix $[k]$ is the element stiffness matrix.

For *nonlinear analysis*, the relation (4.22) should be written in an incremental form; in addition, an iterative procedure should be adopted in the solution of the structure equilibrium equations. Nonlinearity arises from material inelasticity that is irreversible and load-path dependent, large deformation effects (geometric nonlinearity), or both. Therefore, an incremental formulation is necessary in order to implement the material behavior while iterations are essential in order to update geometry. Hence, the relation (4.22) will have the form

$$\{\Delta P\} = [k_t]\{\Delta q\} \tag{4.24}$$

where
    $\{\Delta P\}$ and $\{\Delta q\}$ are the incremental load and displacement vectors, respectively
    $[k_t]$ is the tangent stiffness matrix

The tangent stiffness matrix of a nonlinear material is derived from the incremental constitutive relations (Chen and Han, 1988), as illustrated in Chapter 3:

$$d\sigma_{ij} = C_{ijkl}^{ep} \, d\varepsilon_{kl} \tag{4.25}$$

where
    $d\sigma_{ij}$ and $d\varepsilon_{kl}$ are the stress and strain increment tensors, respectively
    C_{ijkl}^{ep} is the elastic–plastic tensor of tangent moduli

Thus,

$$[k_t] = \int [B]^t [C_{ijkl}^{ep}][B] dV \tag{4.26}$$

4.2.4 ILLUSTRATIVE EXAMPLES

The cantilever beams shown in Figure 4.2, with span to depth ratios, $L/H = 10$ and $L/H = 5$, have been analyzed using the FEM with two options of material behavior: (1) elastic–perfectly plastic (with Young's modulus, $E = 203.9 \times 10^3$ MPa and yield stress, $f_y = 240$ MPa) and (2) elastic–linear work-hardening model (with E and f_y as in option (1), ultimate stress, $f_u = 370$ MPa and ultimate strain, $\varepsilon_u = 0.12$). For both material models, von Mises failure criterion has been adopted. Geometric nonlinearity has been accounted for in the analysis. The analysis has been performed for two cases: (1) beam with no opening and (2) beam with opening as illustrated in the figure.

    The results of the analysis are illustrated in the form of load–displacement relation, Figure 4.3, and strain profile at section 1, near the support, Table 4.1.

    The strain profiles of the beam with $L/H = 10$ and with no opening are almost linear for the load up to the start of the yield and well near failure for either material option (elastic–perfectly plastic or elastic–linear work-hardening model). This validates Bernoulli's hypothesis, a drastic simplification that leads to the simple beam theory. On the other hand, for the beam with opening, Bernoulli's hypothesis is not valid, even before yielding, due to discontinuity. Note that the nonlinearity of strain profile near yield is higher than that near failure. This will be explained later.

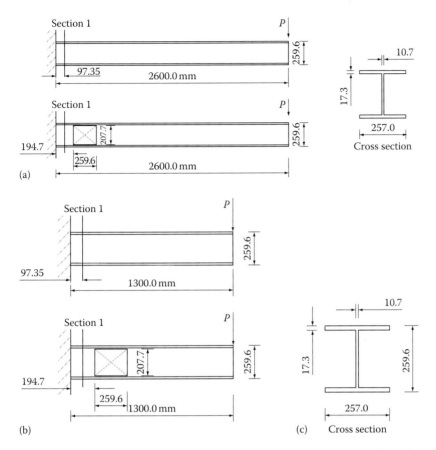

FIGURE 4.2 Examples of cantilever beams: (a) longitudinal sections of cantilever beams with $L/H = 10$; (b) longitudinal sections of cantilever beams with $L/H = 5$; and (c) cross section.

The strain profiles of the beam with $L/H = 5$ and with no opening are almost linear for the load up to the start of yield but not beyond yield. The nonlinearity of strain profile beyond yield is a direct result of the significant shear contribution in such a case. As for the beam with opening, the strain distribution is highly nonlinear as a result of discontinuity caused by the opening.

For the beam with $L/H = 5$ and with opening, the nonlinearity of the strain profile at yield is higher than that of the strain profile near failure. This behavior occurs because the shear strain contribution before yield has a significant effect on beam deformation. Once the section has plasticized, the increase in shear level occurs at a smaller rate compared with the displacement, where the longitudinal strains' contribution has a higher effect on beam deformation. The same behavior could also be observed where opening exists for $L/H = 10$ due to the fact that the opening causes nonlinear strain distribution in the region where strains are measured.

The bazaar strain profile at yield for both material models where opening exists for $L/H = 5$ occurs due to the aforementioned behavior in addition to the nonlinearity imposed by the existence of the opening as described. Once yielding occurs, the

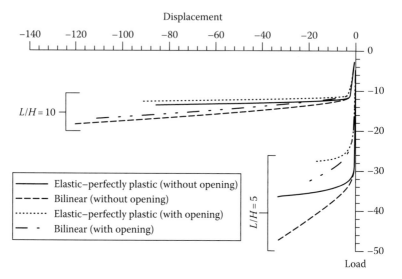

FIGURE 4.3 Load–displacement relation of the example cantilever beams.

shear strain contribution is reduced versus the normal strain contribution and the strain diagrams show less nonlinearity.

With the FEM it is possible to solve these problems with different options: material inelasticity, geometric nonlinearity, imperfection, residual stresses, etc. However, this may not be possible by other means. For instance, simplified solutions may be obtained for beams without opening but not for beams with opening; the capacity of beams with $L/H = 10$ and with no opening can be estimated using the plastic-hinge concept. The FEM enables us to perform sophisticated analysis and accounts for various factors in any type of structure including the regions of discontinuity. However, in these regions, mesh refinement and adjustment are the key for obtaining accurate results.

4.3 APPLICATION FOR STRUCTURAL STEEL MEMBER DESIGN

4.3.1 Column Design Equations

The axial capacity of an axially loaded column can be determined from either an eigenvalue approach or the load-deflection concept. The use of load-deflection analysis of a geometrically imperfect column, which is the practical case, allows tracing the load-deflection response of the member from the start of loading to failure. In this analysis, geometric and material nonlinearity are accounted for explicitly, and, hence, the assessment of column strength is more realistic.

Based on the load-deflection concept (Johnston, 1976), the 1986 American Institute of Steel Construction-Load and Resistance Factor Design (AISC-LRFD) column curve was developed, which has the form

$$\frac{P_n}{P_y} = (0.658)^{\lambda_c^2} \quad \text{for } \lambda_c \leq 1.5 \qquad (4.27a)$$

TABLE 4.1
Strain Distribution at Section 1 of the Example Cantilever Beams (Figure 4.2)

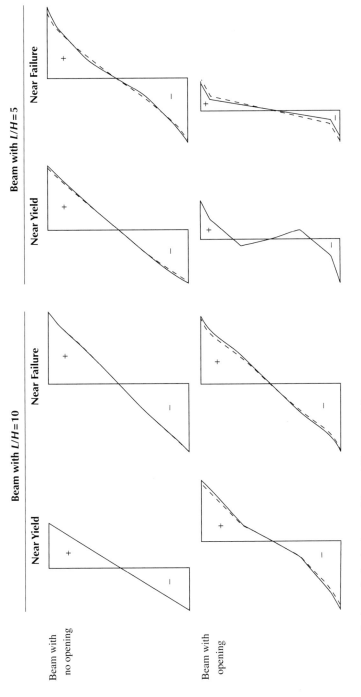

Note: — elastic–perfectly plastic material; - - - bilinear material.

$$\frac{P_n}{P_y} = \frac{0.877}{\lambda_c^2} \quad \text{for } \lambda_c > 1.5 \tag{4.27b}$$

where

$$\lambda_c = \sqrt{\frac{F_y}{F_E}} = \frac{KL}{r}\sqrt{\frac{F_y}{\pi^2 E}} \tag{4.27c}$$

where F_E is the critical flexural buckling stress and is given by $\pi^2 E/(KL/r)^2$. Equation 4.27 was obtained from the curve-fitting of available analytical and experimental data on column strength as well as calibration against Allowable Stress Design (ASD) curve with a live load to dead load ratio of 3. The equation is plotted in Figure 4.4 along with the Column Research Council (CRC) column curve and American Institute of Steel Construction-Allowable Stress Design (AISC-ASD) curve, which were derived from eigenvalue solutions.

Based on rigorous FE analysis of centrally loaded columns with geometric and material imperfections (Bjorhovde, 1972), the variation of maximum strength of steel columns is illustrated in Figure 4.5, where 112 computed maximum strength curves are plotted for a wide range of column shapes and types. Each curve is based on an actual residual stress distribution and an assumed initial out-of-straightness at a mid-height of 0.001 of column length (L/1000). The spread or scatter of computed critical load curves of initially straight columns would be much greater for the same 112 shapes. In order to reduce uncertainty about column strength three subgroups are defined, each of which is represented by a single "average" curve. The three curves, shown in Figures 4.6 through 4.8 and given in Table 4.2, are recommended

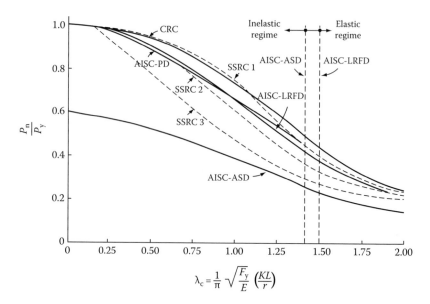

FIGURE 4.4 AISC and SSRC column design curves.

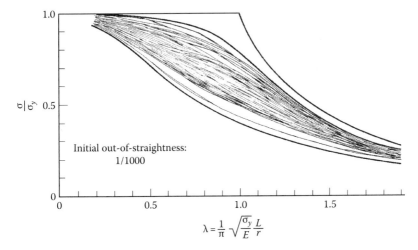

FIGURE 4.5 Maximum strength curves for a number of different column types computed by Bjorhovde (1972).

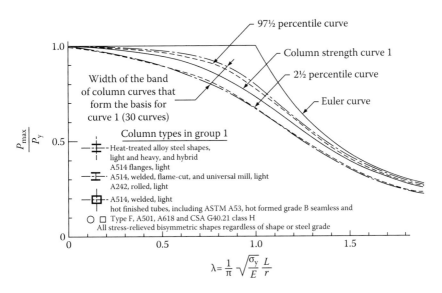

FIGURE 4.6 SSRC column strength curve number 1 for structural steel.

by the Structural Stability Research Council (SSRC) to be used for representing the basic column strength. For comparison, the three curves are plotted in Figure 4.4 along with the AISC-LRFD, CRC, and AISC-ASD curves. From the obtained results, it can be concluded that the FE analysis can account for any variables into consideration; however, a reliability analysis of the obtained results is essential in order to derive design equations.

Based on an extensive analytical and probabilistic study of the experimental strength of centrally loaded columns, the European Convention for Constructional

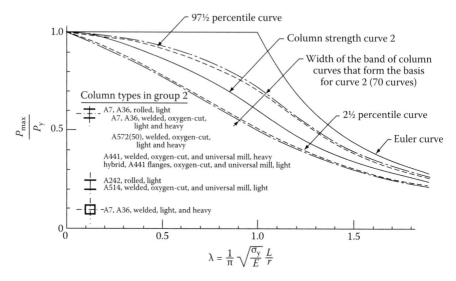

FIGURE 4.7 SSRC column strength curve number 2 for structural steel.

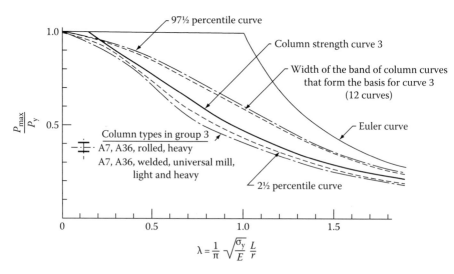

FIGURE 4.8 SSRC column strength curve number 3 for structural steel.

Steel (ECCS) adopted a set of five column curves to represent the basic column strength. These curves are illustrated in Figure 4.9.

4.3.2 BEAM DESIGN EQUATIONS

Based on the concept of limit states, three types of strength limits can be identified for flexural members: (1) formation of plastic hinge, (2) lateral torsional instability, and (3) local instability (flange local buckling or web local buckling). The moment

---

TABLE 4.2

SSRC Multiple Column Curves

SSRC curve 1

$$\frac{P_n}{P_y} = \begin{cases} 1 \,(\text{yield level}) & (0 \le \lambda_c \le 0.15) \\ 0.990 + 0.122\lambda_c - 0.367\lambda_c^2 & (0.15 \le \lambda_c \le 1.2) \\ 0.051 + 0.801\lambda_c^{-2} & (1.2 \le \lambda_c \le 1.8) \\ 0.008 + 0.942\lambda_c^{-2} & (1.8 \le \lambda_c \le 2.8) \\ \lambda_c^{-2} \,(\text{Euler buckling}) & (\lambda_c \ge 2.8) \end{cases}$$

SSRC curve 2

$$\frac{P_n}{P_y} = \begin{cases} 1 \,(\text{yield level}) & (0 \le \lambda_c \le 0.15) \\ 1.035 + 0.202\lambda_c - 0.222\lambda_c^2 & (0.15 \le \lambda_c \le 1.0) \\ -0.111 + 0.636\lambda_c^{-1} + 0.087\lambda_c^{-2} & (1.0 \le \lambda_c \le 2.0) \\ 0.009 + 0.877\lambda_c^{-2} & (2.0 \le \lambda_c \le 3.6) \\ \lambda_c^{-2} \,(\text{Euler buckling}) & (\lambda_c \ge 3.6) \end{cases}$$

SSRC curve 3

$$\frac{P_n}{P_y} = \begin{cases} 1 \,(\text{yield level}) & (0 \le \lambda_c \le 0.15) \\ 1.093 - 0.622\lambda_c & (0.15 \le \lambda_c \le 0.8) \\ -0.128 + 0.707\lambda_c^{-1} - 0.102\lambda_c^{-2} & (0.8 \le \lambda_c \le 2.2) \\ 0.008 + 0.792\lambda_c^{-2} & (2.2 \le \lambda_c \le 5.0) \\ \lambda_c^{-2} \,(\text{Euler buckling}) & (\lambda_c \ge 5.0) \end{cases}$$

---

capacity, according to the LRFD, is obtained as the lowest value considering all these limit states.

For $\lambda \le \lambda_p$

For limit state of formation of plastic hinge,

$$M_n = M_p \tag{4.28}$$

For $\lambda_p < \lambda \le \lambda_r$

For limit state of inelastic lateral torsional buckling,

$$M_n = C_b \left[M_p - (M_p - M_r) \left(\frac{\lambda - \lambda_p}{\lambda_r - \lambda_p} \right) \right] \le M_p \tag{4.29a}$$

where C_b is the moment enhancement factor of the beam, which accounts for the moment gradient. Defining the ratio of the numerically smaller to larger end

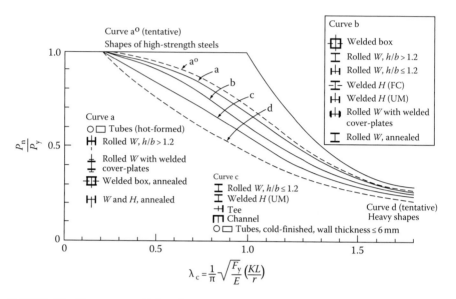

FIGURE 4.9 European multiple column curves.

moments of the beam as M_A/M_B, which is positive if the beam is in double curvature and negative otherwise,

$$C_b = 1.75 + 1.05\left(\frac{M_A}{M_B}\right) + 0.3\left(\frac{M_A}{M_B}\right)^2 \leq 2.3 \qquad (4.29b)$$

For limit state of flange and web buckling,

$$M_n = M_p - (M_p - M_r)\left(\frac{\lambda - \lambda_p}{\lambda_r - \lambda_p}\right) \qquad (4.30)$$

For $\lambda > \lambda_r$

For limit state of elastic lateral torsional buckling,

$$M_n = SF_{cr} \qquad (4.31)$$

where S is the section modulus.

The slenderness parameters λ, λ_p, and λ_r, the limiting buckling moment M_r, and the elastic buckling stress F_{cr} are defined as follows for an I-section.

4.3.2.1 For Strong Axis Bending

For limit state of lateral torsional buckling,

$$\lambda = \frac{L_b}{r_y} \tag{4.32}$$

where L_b is the laterally unbraced length of beam

$$\lambda_p = \frac{300}{\sqrt{F_{yf}}} \tag{4.33}$$

$$\lambda_r = \frac{X_1}{(F_{yw} - F_r)} \sqrt{\left\{ 1 + \sqrt{[1 + X_2(F_{yw} - F_r)^2]} \right\}} \tag{4.34}$$

where

$$X_1 = \frac{\pi}{S_x} \sqrt{\frac{EGJA}{2}}, \quad X_2 = 4 \frac{C_w}{I_y} \left(\frac{S_x}{GJ} \right)^2 \tag{4.35}$$

where

F_{yf} is the flange yield stress
F_{yw} is the web yield stress
F_r is the compressive residual stress in the flange
S_x is the section modulus about the x-axis
I_y is the moment of inertia about the y-axis
r_y is the radius of gyration about the y-axis
E is the modulus of elasticity
G is the shear modulus
J is the torsion constant
C_w is the warping constant

$$M_r = (F_{yw} - F_r)S_x \tag{4.36}$$

$$F_{cr} = \frac{C_b X_1 \sqrt{2}}{\lambda} \sqrt{1 + \frac{X_1^2 X_2}{2\lambda^2}} \tag{4.37}$$

For limit state of flange local buckling,

$$\lambda = \frac{b_f}{2t_f} \tag{4.38}$$

$$\lambda_p = \frac{65}{\sqrt{F_{yf}}} \tag{4.39}$$

$$\lambda_r = \begin{cases} \dfrac{141}{\sqrt{F_{yw} - 10}}, & \text{for rolled shapes} \\[4mm] \dfrac{106}{\sqrt{F_{yw} - 16.5}}, & \text{for welded shapes} \end{cases} \qquad (4.40)$$

$$M_r = (F_{yw} - F_r)S_x \qquad (4.41)$$

$$F_{cr} = \begin{cases} \dfrac{20{,}000}{\lambda^2}, & \text{for rolled shapes} \\[4mm] \dfrac{11{,}200}{\lambda^2}, & \text{for welded shapes} \end{cases} \qquad (4.42)$$

For limit state of web local buckling,

$$\lambda = \frac{h_c}{t_w} \qquad (4.43)$$

$$\lambda_p = \frac{640}{\sqrt{F_{yf}}} \qquad (4.44)$$

$$\lambda_r = \frac{970}{\sqrt{F_{yf}}} \qquad (4.45)$$

$$M_r = F_{yf}S_x \qquad (4.46)$$

In Equation 4.43, h_c is the clear distance between flanges minus the fillet or corner radius at each flange.

4.3.2.2 For Weak Axis Bending

For weak axis bending, the only applicable limit states are the formation of plastic hinge and web local buckling. Thus, $M_n = M_p$ if $\lambda \le \lambda_p$. If $\lambda > \lambda_p$, M_n is obtained from Equation 4.30 with λ, λ_p, and λ_r as defined by Equations 4.43 through 4.45 and $M_r = F_{yw}S_y$.

4.3.3 Beam-Column Design Equations

Based on the analytical solution of 82 inelastic beam-columns (Chen and Lui, 1991), the 1986 AISC-LRFD Specifications recommends the following interaction equations for the design of beam-columns:

For $P_u/\phi_c P_n \geq 0.2$

$$\frac{P_u}{\phi_c P_n} + \frac{8}{9}\left(\frac{M_{ux}}{\phi_b M_{nx}} + \frac{M_{uy}}{\phi_b M_{ny}}\right) \leq 1.0 \qquad (4.47a)$$

For $P_u/\phi_c P_n < 0.2$

$$\frac{P_u}{2\phi_c P_n} + \frac{M_{ux}}{\phi_b M_{nx}} + \frac{M_{uy}}{\phi_b M_{ny}} \leq 1.0 \qquad (4.47b)$$

where
  P_n is the axial compression capacity of the axially loaded column, Equation 4.2.1
  M_{nx} and M_{ny} are the moment resisting capacities of the laterally unsupported beam bent about the x- and y-axes, respectively
  ϕ_c is the column resistance factor (= 0.85)
  ϕ_b is the beam resistance factor (= 0.90)
  P_u is the required axial strength
  M_{ux} and M_{uy} are the required flexural strengths of the members about the x- and y-axes, respectively, calculated as

$$M_u = B_1 M_{nt} + B_2 M_{lt} \qquad (4.48)$$

where
  M_{nt} is the moment in the member assuming that there is no lateral translation in the frame, obtained from first-order elastic analysis (Figure 4.10b)
  M_{lt} is the moment in the member as a result of lateral translation of the frame only, obtained from first-order elastic analysis (Figure 4.10c)
  B_1 and B_2 are the $P-\delta$ and $P-\Delta$ effects (Section 2.4.5)

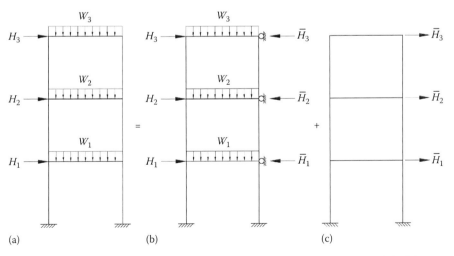

FIGURE 4.10 Determination of M_{nt} and M_{lt} for beam-column design: (a) original frame; (b) nonsway frame analysis for M_{nt}; and (c) sway frame analysis for M_{lt}.

The C_m value used is the same as that in ASD except that the limit $C_m \geq 0.4$ for members braced against joint translation and not subjected to transverse loading is removed.

4.4 APPLICATION FOR STRUCTURAL SYSTEM DESIGN

4.4.1 ADVANCED ANALYSIS FOR STEEL DESIGN

The FEM was a logical extension of the concept of generalized stress and generalized strain. Thus, with the development of FE formulation and with structure discretization, the era of FEM started. It has become possible to analyze a structure member or an entire structure system with any geometry, any loading, and any boundary condition. Moreover, with incremental formulation of the system equilibrium equations and the development of powerful solution strategies in addition to the advancement of material modeling, it has become possible to perform advanced structural analysis, for example, plastic-zone analysis.

In the plastic-zone analysis method of steel frames, members are discretized into FEs, and the cross section of each FE is subdivided into many fibers as shown in Figure 4.11. The deflection at each division point along a member is obtained by numerical integration. The incremental load-deflection response at each loading step, which updates the geometry, captures the second-order effects. The residual stress in each fiber is assumed constant since the fibers are very small. The stress state at each fiber can be explicitly traced, so the gradual spread of yielding can be captured.

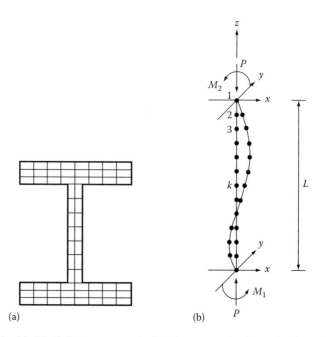

(a)   (b)

FIGURE 4.11 Model of plastic-zone analysis: (a) cross-section discretization and (b) member division.

The plastic-zone analysis eliminates the need for separate member capacity checks since it explicitly accounts for second-order effects, spread of plasticity, and residual stresses. As a result, the plastic-zone solution is known as an "exact solution."

The plastic-zone analysis may be performed by using either three-dimensional finite shell elements or the beam-column theory (Chen and Kim, 1997). In the three-dimensional finite shell elements, the elastic constitutive matrix in the usual incremental stress–strain relations is replaced by an elastic–plastic constitutive matrix when yielding is detected. Based on the deformation theory of plasticity, the effects of combined normal and shear stresses may be accounted for. This analysis requires modeling of structures using a large number of three-dimensional finite shell elements and numerical integration for the evaluation of the elastic–plastic stiffness matrix. The three-dimensional spread-of-plasticity analysis, when combined with second-order theory, which deals with frame stability, is computationally intensive. Therefore, it is best suited for analyzing small-scale structures and providing detailed solutions for local instability and yielding behavior of members, if required. Moreover, it can be used to provide solutions for benchmark problems; that is, it can be used to provide benchmark models, as illustrated in Chapter 6, or replace physical testing.

4.4.2 FE Analysis of Offshore Concrete Structures

In spite of the success of the FEM when applied to steel structures, there are still several open questions and some difficulties in the prediction of the ultimate behavior of reinforced concrete structures. The FEM is most successful when dealing with composite structures with a perfect bond between steel and concrete, without tensile cracks and without the localization of discrete failure zones or strain softening. In the following, we shall show the experience learned from the application of FEM to offshore structures in earlier years as an illustrative example.

4.4.2.1 Research behind the Success of the Offshore Structures

The offshore concrete structures constructed in the 1970s for the North Sea oil development were analyzed extensively using the FEM. Some of the highlights of the analysis process are summarized in the following (Chen, 2000):

- Solve 100,000 simultaneous equations
- Designed for a 30 m wave with the platform in a 300 m or 1000 ft deep water
- Consider 25,000 load combinations
- Use supercomputer for computing
- Assume the material to be linearly elastic
- Cost $7 M to develop the computer program
- Require 250 engineers to input the data

4.4.2.2 Failure Experience, the Problem

Although applying the FE in this mega project was successful, the offshore concrete platform Sleipner A was totally lost in the North Sea during the installation process.

Failure was due to poor detailing essentially reflecting deficiencies of basic nature in codes as well as in practice (Schlaich and Reineck, 1993). The reasons for the failure of the analysis are as follows:

- It was not considered that the concrete will crack after overloading and redistribution of the stresses.
- The anchorage of the reinforcing bars was found to be inadequate in the tension zone after the crack of concrete.
- Failure cost exceeded $0.5B for the structure and $1B for the overall economy.
- The computed shear force by the FEM was about 60% of simple beam hand calculations.

4.4.2.3   Lessons Learned

The subsequent analysis and design for a successful construction of the platform considered the following improvements:

- Improve the model on strength and deformation of reinforced concrete element under all possible load combinations and torsion.
- Carry out large-scale element tests for both strength and fatigue.
- Conduct biaxial compression/tension tests; compressive strength increases by lateral compression and decreases by every cycle of tension.
- Much more steel is necessary in the shells.
- Use concrete with a slump of 260 mm (10 in.) instead of 120 mm (4.7 in.) to get through.
- Shells are too heavy for installation; use light-weight concrete to reduce weight.

4.4.2.4   Concluding Remarks
- Engineers need to develop a good material model for a heavily reinforced concrete plate element.
- Need to carry out simple hand calculations to check the computer solutions.
- Need experienced engineers to do hand calculation check.
- Need to consider partial failure analysis, like cracks, to see possible redistribution.

4.4.3   A GLANCE TO THE FUTURE

With the exponential growth of computer speed and capacity the implementation of model-based design is emerging. Model-based simulation has been implemented in aerospace for system design and manufacturing of *Boeing 777*, in auto industry for component design for crashworthiness and for system design of the next generation

Space Telescope (Hubble II). This emerging area of model-based simulation in structural engineering is illustrated in Chapter 7.

4.5 LOAD AND RESISTANCE FACTOR DESIGN FOR STRUCTURAL STEEL BUILDINGS

4.5.1 RELIABILITY-BASED LRFD CODE AS A START

Design methods have developed since the era of elasticity from working-stress design to plastic design to limit-state design or load-resistance-factor design. The FE has not only provided remarkable flexibility or increased capability of structural analysis but also influenced the design methodology. With the aid of the FE, numerous data related to the performance of different structural elements can be generated, for example, the case of column under axial force. On one side, the analysis of the generated data in order to obtain design information necessitated the implementation of stochastic analysis. On the other, the uncertainty about environmental conditions requires stochastic models for realistic assessment and implementation of these conditions in design. With regard to structure safety, for example, the introduction of partial safety factors was a natural development for more rational design.

The load that a structure may experience during its design life, S, and the strength, R, involve degrees of uncertainty and randomness. Structure safety can be identified in a simple manner by S and R; that is, if $R > S$ the structure is safe, and if $R < S$ the structure will fail. For simplicity, S and R can be assumed independent, which is a correct assumption for the usual cases of static and quasi-static loading. The probability density functions of S and R are defined as $f(S)$ and $g(R)$, respectively, and are illustrated in Figure 4.12. The probability of failure, probability that $R < S$, is

$$P_F = \int_0^\infty f(S) \left[\int_0^S g(R)\, dR \right] dS \qquad (4.49)$$

and the structure reliability is $(1 - P_F)$.

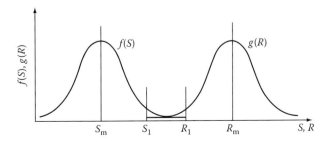

FIGURE 4.12   Load and resistance probability functions.

In order to ensure an acceptable structure reliability, the structure has to be designed for load greater than the nominal value and strength smaller than the nominal value, that is,

$$\phi R_n \geq \gamma S_n \qquad (4.50)$$

where
   ϕ is the strength-reduction factor and is less than 1.0
   γ is the load factor and is greater than 1.0
   R_n and S_n are the nominal strength and nominal load, respectively

For load combination, Equation 4.50 takes the form

$$\phi R_n \geq \sum_{i=1}^{m} \gamma_i S_{ni} \qquad (4.51)$$

where
   i is the load type (e.g., dead load or live load)
   m is the number of load types
   γ_i is load factor corresponding to type i

4.5.2 LOAD FACTORS

LRFD uses different load factors for each load type in order to reflect the degree of uncertainty of this load. The load factors were developed from statistical analysis, and the strength reduction factors were initially derived from calibration with the ASD. Nevertheless, these factors were decided in order to guarantee that the design satisfies certain safety (or reliability) of structural members. In other words, the design guarantees that the probability of exceeding a limit state of the structural member (e.g., yielding, fracture, or buckling) is less than certain limit. The factors and load combinations used by LRFD are summarized in Table 4.3.

4.5.3 RESISTANCE FACTORS

For safety, all codes adopted the partial safety factors with difference in the application of the concept. In the LRFD or the ACI, the partial strength reduction factors are applied to the nominal strength, for example, the nominal moment or nominal shear. In the Eurocode, these factors are applied to the material strength, for example, concrete characteristic strength or steel yield stress, and the reduced material strengths are used in the calculation of the design of section or member strength. The strength-reduction factors used by LRFD are given in Table 4.4.

In ASD, the stresses in a structure under working or service loads should not exceed the designated allowable values. The allowable values are obtained by

TABLE 4.3

Load Factors and Load Combinations

$1.4D$

$1.2D + 1.6L + 0.5(L_r \text{ or } S \text{ or } R)$

$1.2D + 1.6(L_r \text{ or } S \text{ or } R) + 0.5(L \text{ or } 0.8W)$

$1.2D + 1.3W + 0.5L + 0.5(L_r \text{ or } S \text{ or } R)$

$1.2D \pm 1.0E + 0.5L + 0.2S$

$0.9D \pm (1.3W \text{ or } 1.0E)$

D, dead load; L, live load; L_r, roof load; W, wind
load; S, snow load; E, earthquake load; R, nominal load due to initial rainwater on ice exclusive
of the ponding contribution.

TABLE 4.4

Strength-Reduction Factors

Member type and limit state	ϕ
Tension member, limit state: yielding	0.90
Tension member, limit state: fracture	0.75
Pin-connected member, limit state: tension	0.75
Pin-connected member, limit state: shear	0.75
Pin-connected member, limit state: bearing	0.75
Column; compression	0.85
Beams; flexure	0.90
High-strength bolts, limit state: tension	0.75
High-strength bolts, limit state: shear	0.75

dividing the yield stress or material strength by a factor of safety. The general form
of this concept is

$$\frac{R_n}{F.S.} \geq \sum_{i=1}^{m} S_{ni} \qquad (4.52)$$

where $F.S.$ is the factor of safety, which is greater than 1.0 (e.g., 1.5 for beams and
1.67 for tension members). The left-hand side of Equation 4.52 represents the allowable stress of the structural member or component under a given loading condition
(e.g., tension, compression, and bending). The right-hand side of the equation represents the combined stress produced by various load combinations (e.g., dead, live,
and wind). It is noted that the factor of safety is applied only to the strength term,
and safety is evaluated at the service load, with all load types given the same weight
regardless of their uncertainty.

4.5.4   PERFORMANCE-BASED DESIGN AS A CURRENT PROGRESS

During its design life, a structure should fulfill its function. This dictates the consideration of all possible environmental conditions, misuse or possible change of use, possible deterioration of structure components, etc. Along this line came performance-based design, which is an engineering approach to structural design that is based on agreed-upon performance goals and objectives. Examples of these are serviceability performance, structure response against earthquakes, and fire resistance of structures. In order to achieve any of these goals, performance criteria are established and are required to be satisfied by the structure design.

Serviceability performance is a state in which the function of a building, its appearance, maintainability, durability, and comfort of its occupants are preserved under normal usage. Limiting values of structural behavior for serviceability (e.g., maximum deflection, camber, drift, acceleration, vibration, wind-induced motion, expansion, and contraction) are decided with due regard to the intended function of the structure.

Design for earthquakes is governed by the desired structure performance against an earthquake. For instance, for minor and moderate earthquakes, the structure is required to respond elastically. On the other hand, for major earthquakes, the structure is required to perform in the inelastic or plastic range with some minimum requirement of energy dissipation, however, with plastic deformations desired in certain locations.

Structural components, members, and building systems shall be designed so as to maintain their load-bearing function during the design-basis fire and to satisfy other performance requirements specified for building occupancy. Three limit states existing for elements serving as fire barriers (compartment walls and floors) shall be considered in design. These states are (1) heat transmission leading to unacceptable temperature rise on the unexposed surface, (2) breach of barrier due to cracking or loss of integrity, and (3) loss of load-bearing capacity. Deterioration shall be applied where the means of providing structural fire resistance, or the design criteria for fire barriers, require consideration of the deterioration of load-carrying structure.

With the development and advancement of the FEM, a performance-based design has become within the capability of the designer. Either simplified analysis or rigorous analysis can be performed in order to do the necessary design adjustments, which meet the designated design criterion.

4.6   HISTORICAL SKETCH

The FEM is a computer-based numerical analysis technique for obtaining approximate solution to a wide variety of engineering problems. It was originally developed to study stresses in complex airframe structures, but it has now been extended and applied to a variety of continuum mechanics problems including structural engineering.

The term *FEM* was first used by Clough in his 1960 paper on plane elasticity problems. However, the concept of FE analysis could trace back much early depending on whether one asks an applied mathematician, a physicist, or an engineer. In the field of applied mathematics, Courant in 1943 was the first to use an assemblage of triangular elements and the principle of minimum potential energy to study the St. Venant torsion problem. In the meantime, physicists were also busy developing similar ideas. For example, the work of Prager and Synge (1947) led to the development of the hypercircle method, which can be applied to continuum problems in much the same way as FE is applied.

Faced with increasing complex problems in aerospace structures, structural engineers with their physical intuition used the truss concept to treat a structure as an assemblage of a finite number of interconnected nodal points with rod-like structural elements to replace the real structure. Under this discretization process, the problem has now reduced to that of the conventional old structural analysis. As a result of this success, the seed to FE techniques began to germinate in the structural engineering community.

With the advent of digital computers at that time, the actual solution of plane stress problems by means of triangular elements whose properties were derived from the theory of elasticity was first given in the now classical paper of Turner et al. (1956). In a 1980 paper, Clough gives his personal account of the origins of the method describing the sequence of the events. In the years after 1960, the FEM was received widely in all fields of engineering. By 1972, the FEM had become the most active field of interest in the numerical solution of continuum problems. It remains the dominant method in developing advanced analysis methods and design specifications for structural engineering community today.

REFERENCES

Bjorhovde, R., 1972, Deterministic and probabilistic approaches to the strength of steel columns, PhD dissertation, Lehigh University, Bethlehem, PA.

Chen, W. F., 2000, Plasticity, limit analysis and structural design, *International Journal of Solids and Structures*, 37, 81–92.

Chen, W. F. and Han, D. J., 1988, *Plasticity for Structural Engineers*, Springer-Verlag, New York.

Chen, W. F. and Kim, S. E., 1997, *LRFD Steel Design Using Advanced Analysis*, CRC Press, Boca Raton, FL.

Chen, W. F. and Lui, E. M., 1991, *Stability Design of Steel Frames*, CRC Press, Boca Raton, FL.

Clough, R. W., 1960, The finite element method in plane stress analysis, *Proceedings of the Second ASCE Conference on Electronic Computation*, Pittsburgh, PA, September 8–9.

Clough, R. W., 1980, The finite element method after twenty-five years: A personal view, *Computers & Structures*, 12, 361–370.

Courant, R., 1943, Variational methods for the solutions of problems of equilibrium and vibrations, *Bulletin of the American Mathematical Society*, 49, 1–23.

Desai, C. S. and Abel, J. F., 1972, *Introduction to the Finite Element Method: A Numerical Method for Engineering Analysis*, Van Nostrand Reinholdt Company, New York.

Johnston, B. J., 1976, *Guide to Stability Design Criteria for Metal Structures*, 3rd edn., John Wiley & Sons, Inc., New York.

Prager, W. and Synge, J. L., 1947, Approximation in elasticity based on the concept of function space, *Quarterly of Applied Mathematics*, 5, 241–269.

Schlaich, J. and Reineck, K., 1993, Die Ursache für den Totalverlust der Betonplattform Spleiner A, Beton-und Stahlbetonbau 88, H. 1, Ernst & Sohn Verlag für Architektur and technische Wissenchaften, Berlin, 1993.

Turner, M. J., Clough, R. W., Martin, H. C., and Topp, L. C., 1956, Stiffness and deflection analysis of complex structures, *Journal of Aeronautical Sciences*, 23, 9, 805–823, 854.

Yang, T. Y., 1986, *Finite Element Structural Analysis*, Prentice-Hall, Inc., Englewood Cliffs, NJ.

5 Strut-and-Tie Model for Design of Structural Concrete Discontinuity Regions

5.1 INTRODUCTION

The significance of this new conceptual change in the design of reinforced concrete structures can be best described by an observation made in 1984 by Professor MacGregor of Canada:

> One of the most important advances in reinforced concrete design in the next decade will be the extension of plasticity based design procedures to shear, torsion, bearing stresses, and the design of structural discontinuities such as joints and corners. These will have the advantage of allowing a designer to follow the forces through a structure.

The strut-and-tie model (STM) is a logical extension of the truss model and the major difference between the two methods is that the STM is a set of forces in equilibrium but do not form a stable truss system. Thus, the STM is a generalization of the truss model. The truss model has been recognized in academia and practice to be the most reliable tool for the treatment of shear and torsion in structural concrete B-regions. The STM is currently recognized as the most reliable tool for the treatment of D-regions.

The basic concept of the STM is based on the lower-bound theorem of limit analysis of perfect plasticity. It visualizes a truss-like system in the structure or its components to transfer load to the supports where

- Compression forces are resisted by concrete "struts"
- Tensile forces are resisted by steel "ties"
- Struts and ties meet at "nodes"

For best serviceability, the model should follow the elastic flow of forces. The merit of limit analysis in terms of STM procedures to design lies in the fact that engineers can make practical and safe decisions on the detailing of complex structural discontinuities in reinforced concrete on the basis of relatively simple calculations.

The concept of STM has been illustrated and explained in Section 3.5 within the scope of limit analysis applications. In addition, the failure criteria of the STM elements and an illustrative design example have been presented. This chapter is devoted for in-depth understanding of this powerful procedure for engineering practice. The definition of B- and D-regions is elaborated, and basic D-regions along with their appropriate models are illustrated. Some common D-regions, for example, local pressure, beams with dapped end, beams with recess, deep beams with large openings, corner joints, and exterior and interior beam-column connections, are examined and modeled. Finally, a detailed illustrative design example is presented followed by a historical sketch.

5.2   D-REGIONS VERSUS B-REGIONS

5.2.1   INTRODUCTION

A concrete structure can be subdivided into two types of regions based on the strain distribution within a cross section, which is an influential factor in the design approach of these regions. Those regions where Bernoulli's hypothesis of flexure (plane sections before bending remain plane after bending) can be assumed valid are referred to as Bernoulli or bending regions (or simply *B-regions*). The other regions where Bernoulli's hypothesis does not apply are referred to as discontinuity or disturbance regions (or simply *D-regions*).

B-regions have been successfully treated using the truss model. On the other hand, this truss model has been extended and generalized leading to the STM method for the treatment of D-regions. With this, the entire structure is treated in a consistent manner. The concept of STM with its different elements has been introduced in Section 3.5 as a class of lower-bound solutions of limit analysis. The validity and success of the method had been proven in academia and in practice.

D-regions are usually the most critical regions in structural concrete since they are, by nature, most vulnerable to environmental loading conditions. STM as a transparent and translucent tool represents a rational approach to understanding the behavior of such regions. This chapter aims to view the topic of STM from a design standpoint and to deepen the understanding of the behavior and design of D-regions in structural concrete.

5.2.2   B-REGIONS

B-regions are found in plates and beams where the depth is either constant or changes gradually, and loads are continuously distributed. The state of stress at any section of a B-region can be adequately derived from sectional effects (bending, torsion, shear, and normal force).

The solution of uncracked B-regions can be satisfactorily formulated based on the theory of elasticity as in standard mechanics books. On the other hand, if the tensile stresses in B-regions exceed the tensile strength of concrete, the truss model will apply instead of the elasticity-based solutions. In addition to the truss model, codes of practice (ACI 318-08, Eurocode 2, and ECP 203-2006, among others) permit other standard methods that have passed the test of experiment.

5.2.3 D-Regions

In D-regions, the strain distribution is significantly nonlinear as a result of discontinuity, which results from a sudden change of geometry (geometric discontinuity) or concentrated loads (static discontinuity). Examples of geometric discontinuity are recesses in beams, frame corners, bends, and openings (Figure 5.1a and c). Examples of static discontinuity are the regions of concentrated loads, reactions, and local pressure (such as prestressing anchorage zones; Figure 5.1b and c). Structures such as deep beams, where the strain distribution is significantly nonlinear, are considered as one entire D-region (Figure 5.1b).

Uncracked D-regions can be satisfactorily analyzed based on the theory of elasticity by using, for instance, finite element codes. Nevertheless, this is not the case in most practical applications even under service loads. Once cracks form in a D-region and bond stresses between reinforcement and concrete develop significantly, linear

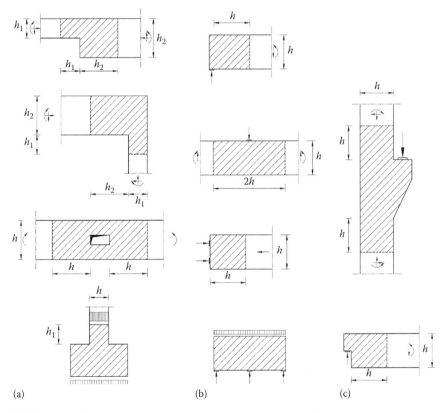

(a)   (b)   (c)

FIGURE 5.1 D-regions (shaded areas) with nonlinear strain distribution due to (a) geometric discontinuity; (b) static discontinuity; and (c) geometric and static discontinuities. (Adapted from Schlaich, J. et al., *J. Prestressed Concr. Inst.*, 32(3), 74, 1987; Schlaich, J. and Schäfer, K., *J. Struct. Eng.*, 69(6), 113, 1991; Schlaich, J. and Schäfer K., The design of structural concrete, *IABSE Workshop*, New Delhi, India, 1993.)

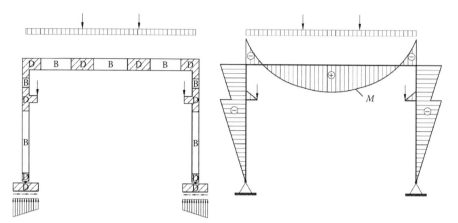

FIGURE 5.2 A frame structure containing a substantial part of B-regions: its statical system and bending moments. (Adapted from Schlaich, J. et al., *J. Prestressed Concr. Inst.*, 32(3), 74, 1987.)

elastic analysis is not applicable any more. On the other hand, a complete nonlinear analysis may turn out to be uneconomical, especially in the early stages of design; besides, it does not help in the development of the right detailing. Moreover, if structure behavior is not precisely simulated, the results may be a cause of poor performance or future failure. With this in mind, the STM method represents the rational approach for the treatment of D-regions (Schlaich and Schäfer, 1991, 1993).

In B-regions, the state of stress may be derived from sectional effects, whereas in D-regions this is not the case. Nevertheless, conventional structural analysis is essential, and with the division of a structure into B- and D-regions, the boundary forces of D-regions can be identified. These boundary forces come from the effect of attached B-regions and other external forces and reactions (Figure 5.2).

5.2.4 DEFINING THE BOUNDARIES OF D-REGIONS

In contrary to D-regions, the stresses and stress trajectories in B-regions are smooth (Figure 5.3). In D-regions, stress intensities decrease rapidly with the distance from the origin of the stress concentration. Such behavior is the key in the identification of B- and D-regions of a structure.

In order to illustrate how the division lines between B- and D-regions are defined, two illustrative examples shown in Figure 5.4 are considered. The common principle is to subdivide the real structure in Figure 5.4(i) into the state of stress, which satisfies Bernoulli's hypothesis, Figure 5.4(ii), and the compensating state of stress, Figure 5.4(iii). Upon applying the principle of St. Venant, Figure 5.5, it is assumed that the nonlinear stresses in Figure 5.4(iii) are negligible at a distance that is approximately equal to the maximum distance between the equilibrating forces themselves. The distance defines the range of the D-regions, Figure 5.4(iv), as illustrated in the examples in Figure 5.4. It should be noted that for most cases of beams, this distance is practically equal to the height of the cross section of adjacent B-regions attached to the D-region.

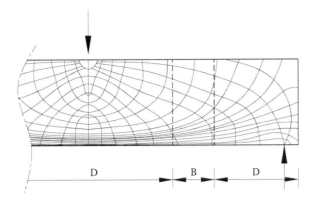

FIGURE 5.3 Stress trajectories in a B-region and near discontinuities (D-regions).

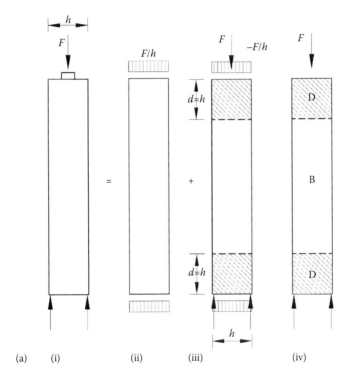

FIGURE 5.4 Two examples of subdivision of structures into their B- and D-regions using St. Venant's principle (a) column or wall with concentrated loads and (b) beam with recess: (i) structure with real load; (ii) loads and support reactions applied in accordance with the Bernoulli hypothesis; (iii) self-equilibrating state of stress; and (iv) real structure with B- and D-regions. (Adapted from Schlaich, J. et al., *J. Prestressed Concr. Inst.*, 32(3), 74, 1987; Schlaich, J. and Schäfer K., The design of structural concrete, *IABSE Workshop*, New Delhi, India, 1993.)

(*continued*)

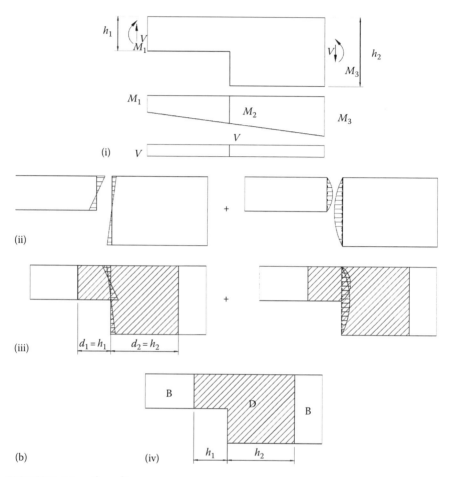

FIGURE 5.4 (continued)

In cracked concrete members, the stiffnesses in different directions may alter as a result of cracking; consequently, the boundaries of D-regions may change as well. Nevertheless, the preceding approach for the determination of the division lines between B- and D-regions, which was based on elastic material behavior, is still applicable. This is due to the fact that the principle of St. Venant itself is not precise and the dividing lines between B- and D-regions serve only as a qualitative aid in the development of STMs.

5.3 STRUT-AND-TIE MODEL AS A SOLUTION

5.3.1 SAFE SOLUTION BASED ON EQUILIBRIUM APPROACH

As explained in Section 3.5, the STM is an idealization of the stress resultants derived from the flow of forces within a region of structural concrete. For the D-region in Figure 5.6a, the boundary forces acting on the region are determined and the stress

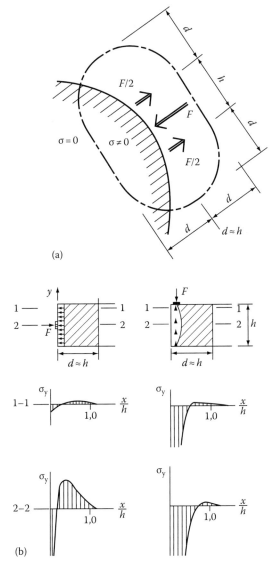

FIGURE 5.5 The principle of St. Venant: (a) zone of a body affected by self-equilibrating forces at the surface and (b) application to a prismatic bar (beam) loaded at one face.

diagrams of these forces are subdivided in such a way that the individual stress resultants on opposite sides of the region correspond in magnitude and can be connected by streamlines that do not cross each other (Figure 5.6b). Then, the flow of forces through the region can be traced using the load-path method or the stress trajectories from linear elastic analysis (Figure 5.6b). The flow of forces, which are smoothly curved, are replaced by polygons as shown in Figure 5.6c, and additional struts and ties are added for equilibrium such as the tie T and the strut C.

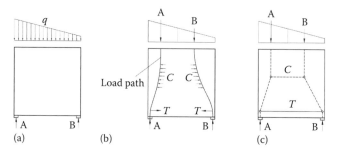

FIGURE 5.6 Development of an STM (Schlaich and Schäfer, 1991): (a) the region and boundary loads; (b) the load paths through the region; and (c) the corresponding STM.

The successful model should satisfy two conditions: equilibrium and failure criteria. Since compatibility conditions are omitted from this method, the obtained solution is a lower-bound or safe estimate. Hence, the STM method always provides a safe solution.

The method of STM with its different elements and failure criteria has been discussed in Section 3.5. Only new complementary thoughts, applications, illustrative points, and examples are presented in the following sections.

5.3.2 BASIC DISCONTINUOUS STRESS FIELDS

D-regions with their respective boundary conditions can be looked at as isolated discontinuous stress fields. Many of these fields are basic (or standard) stress fields and hence have standard STM solutions. These regions are designated by Sclaich and Schäfer (1993) as D_1, D_2, ... D_{12}; nevertheless only D_1 to D_{10} are the important basic D-regions and are illustrated in Figure 5.7.

5.4 SELECTED DISCONTINUOUS STRESS FIELDS

5.4.1 LOCAL PRESSURE

The problem of local pressure, Figure 5.8, is simulated as D_1-region for the case of concentric load, Figure 5.7a, and D_2-region for the case of eccentric load, Figure 5.7b. The amount of necessary transverse reinforcement and its position can be determined from the respective STM. It is noted that this reinforcement has to be closed stirrups for either case. Nevertheless, the location of these stirrups is at the edge of the D-region for the case of eccentric load and moves inward for the case of concentric load. In the case of eccentric load, longitudinal reinforcement is required, as illustrated by the model.

5.4.2 DAPPED BEAM

In order to hang the reaction of a dapped beam, two possible models are shown in Figure 5.9 (Schäfer and El-Metwally, 1994). The first model, Figure 5.9a, illustrates

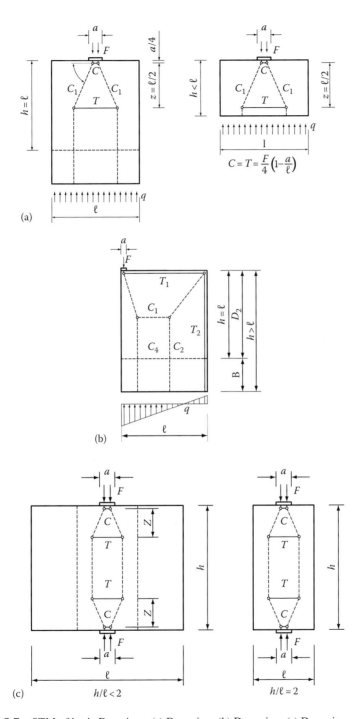

FIGURE 5.7 STM of basic D-regions: (a) D_1-region; (b) D_2-region; (c) D_3-region;

(continued)

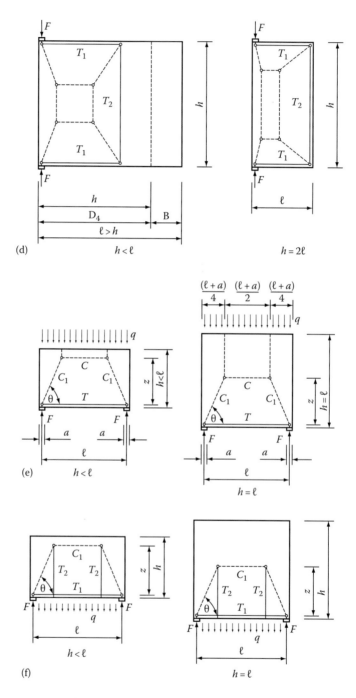

FIGURE 5.7 (continued)   (d) D_4-region; (e) D_5-region; (f) D_6-region;

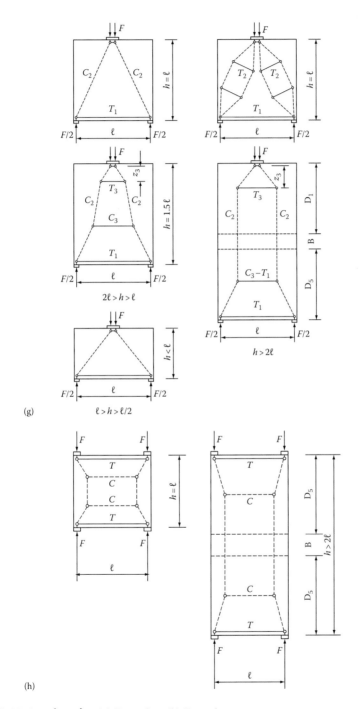

FIGURE 5.7 (continued) (g) D_7-region; (h) D_8-region;

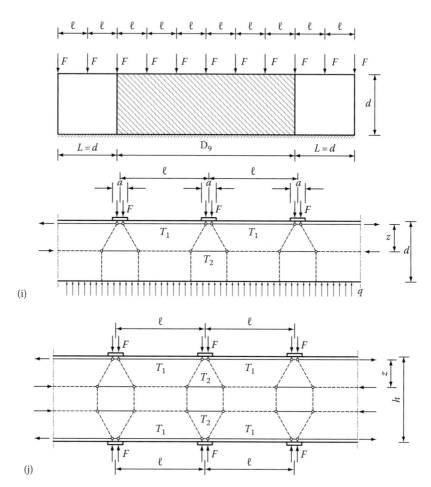

FIGURE 5.7 (continued)   (i) D_9-region; and (j) D_{10}-region. (Modified from Schlaich, J. and Schäfer K., The design of structural concrete, *IABSE Workshop*, New Delhi, India, 1993.)

that in addition to the shear reinforcement, T_3, and the reinforcement necessary for hanging the reaction, T_1, an additional reinforcement is necessary for a safe transfer of the forces within the D-region, that is, tie T_A and the increase in the magnitude of tie T_2 above the shear requirement.

The second alternative, Figure 5.9b, requires lesser reinforcement; nevertheless, anchorage of the inclined reinforcement at the upper nodes may become a problem in case of thick bars. In practice, some reinforcement is always detailed as indicated by the first model, which is necessary for keeping the integrity of the D-region. Hence, though the two presented models are correct, a more efficient detailing can be achieved by a combination of the two models. The combined model shown in Figure 5.9c leads to a more efficient detailing with the inclined reinforcement assigned at most 70% of the load and the first model assigned at least 30% of the load. While the anchoring of the inclined reinforcement is difficult in the second model, this problem

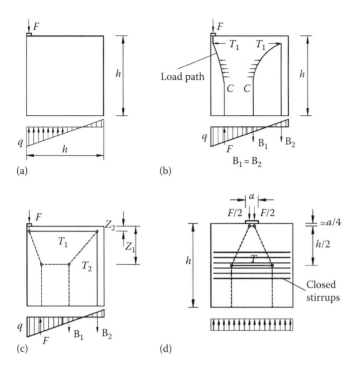

FIGURE 5.8 Local pressure: (a–c) eccentric load and (d) concentric load. (Taken from Schäfer, K. and El-Metwally, S.E., On the role of discontinuity regions detailing in the safety of concrete structures, *Proceedings of the Fifth International Colloquium on Concrete in Developing Countries*, Cairo, Egypt, January 2–6, pp. 43–55, 1994.)

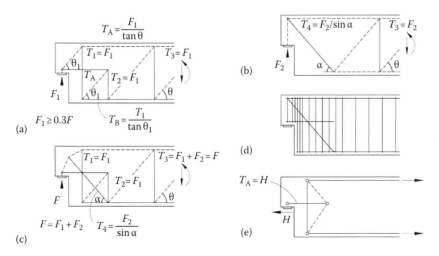

FIGURE 5.9 Dapped beam: (a) STM 1; (b) STM 2; (c) combined model from 1 and 2; (d) reinforcement layout; and (e) STM for a horizontal reaction H.

is relieved by the combined model. The reinforcement layout for the combined model is shown in Figure 5.9d.

If a horizontal reaction force H exists, the model shown in Figure 5.9e can be used for the evaluation of the required additional reinforcement.

5.4.3 Beam with Recess

The beam in Figure 5.10a shows the shaded D-region as a result of the recess shown. For illustration, the D-region is assumed to be subjected to two cases of constant moment (no shear), positive and negative moments, Figure 5.10b, and the moment lever arm on the left end of the D-region is assumed to be one-half of that on the right end.

Upon examination of all possible load paths for either case of end moment, the appropriate STMs can be derived as shown in Figure 5.10c (Schäfer and El-Metwally, 1994). From the obtained models it is noted that anchoring the curtailed longitudinal reinforcement should start at a distance beyond what is

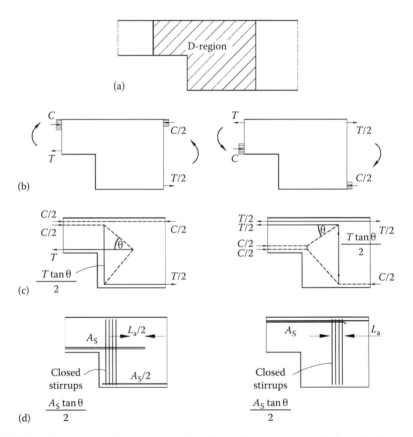

FIGURE 5.10 Beam with recess: (a) D-region; (b) moments applied to the D-region; (c) STMs; and (d) appropriate reinforcement detailing.

required by the sectional design. In addition, for the safety of the D-region, transverse reinforcement in the shape of closed stirrups is necessary in order to carry the transverse tension $1/2\,T\tan\theta$.

5.4.4 WALLS WITH OPENINGS

In Figure 5.11a and b, the STMs of a wall with rectangular opening are given for two cases of uniform compression and uniform tension applied to two opposite boundaries of the wall (Schlaich and Schäfer, 1991). It is obvious from Figure 5.11a that the tie T_1 would require a reinforcement parallel to and near the edge of the opening, which agrees in principle with the normal practice. The quantification of this tie force and hence the amount of the required reinforcement are given by the STM shown in the figure. On the other hand, for a wall under tension the model in Figure 5.11b reveals that reinforcement would be required along the edges parallel to the load to carry tie, T_2, which also agrees with the normal practice. Nevertheless, the reinforcement of tie T_1, which is parallel to the edges perpendicular to the load direction, has to be placed at a distance from these edges and not along the edges, which may not be satisfied in normal practice. In addition, the anchorage of the reinforcement of ties T_2 and T_3 has to be checked considering the additional length due to the lateral shift of the respective forces as shown by the model.

5.4.5 DEEP BEAM WITH ECCENTRIC LARGE OPENING

Due to the applied concentrated load, the deep beam with an eccentric large opening shown in Figure5.12a has the stress trajectories shown in Figure5.12b. Two different STMs are combined together for rational representation of the beam behavior (Schlaich et al., 1987). The simple STM shown in Figure 5.12c for the right part of the beam can be refined in order to account for the transverse stresses in the strut. The refined model is shown in Figure 5.12d, which apparently conforms better with the stress trajectories. As for the left part of the beam, either one of the two models shown in Figure 5.12f and g is justifiable by the stress trajectories. In both

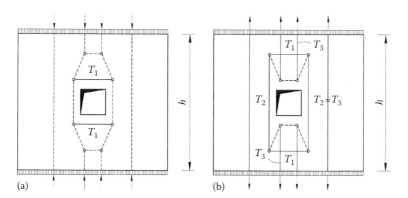

(a) (b)

FIGURE 5.11 Walls with openings: (a) wall under uniform compression and (b) wall under uniform tension.

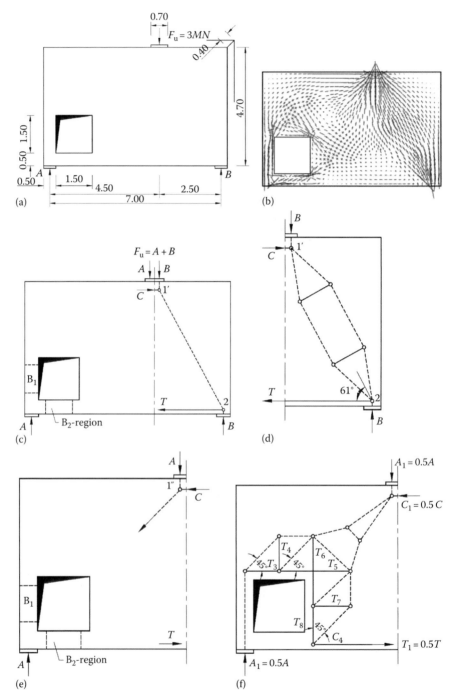

FIGURE 5.12 Deep beam with a large eccentric opening: (a) dimensions and loads; (b) stress trajectories; (c) load path for the right side; (d) complete model for the right side; (e) boundary forces for the left side; (f) model 1 for the left side;

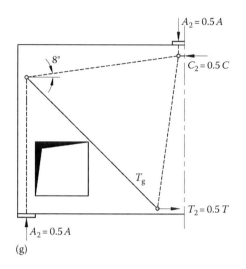

(g)

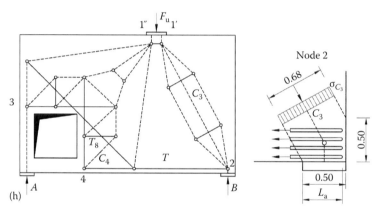

(h)

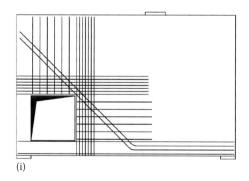

(i)

FIGURE 5.12 (continued) (g) model 2 for the left side; (h) complete STM; and (i) reinforcement. (Adapted from Schlaich, J. et al., *J. Prestressed Concrete Inst.*, 32(3), 74, 1987; Schlaich, J. and Schäfer, K., The design of structural concrete, *IABSE Workshop*, New Delhi, India, 1993.)

models, the support reaction A is transferred vertically until a level above the opening by axial action through the B_1-region rather than horizontally by bending action through the B_2-region. This approach is justified by the fact that the axial stiffness of the B_1-region is much greater than the bending stiffness of the B_2-region.

The second model, Figure 5.12g, requires a lesser reinforcement than the first model, Figure 5.12f. Nevertheless, the inclined reinforcement may have an anchorage problem at the upper left node; in addition, some reinforcement has to be detailed according to the first model for keeping the integrity of the concrete material around the opening. A combination of the two models would lead to the most efficient detailing, for example, by assigning 50% of the load to every model. The combined model is shown in Figure 5.12h with the corresponding tension reinforcement layout shown in Figure 5.12i. Of course, web reinforcement and a minimum reinforcement of the B_2-region would still be required.

5.4.6 KNEE CORNER JOINTS UNDER OPENING MOMENTS

In the design of a reinforced concrete moment-resisting frame structure, the corner geometry can be defined from the dimensions of the structural elements meeting at the joint, beam, and column. The first step in the design of connection using STM method is to identify all forces acting on the D-region (connection). Based on the observed joint behavior and the proposed reinforcement detailing, the appropriate STM can be derived.

The different STMs shown in Figure 5.13a were suggested by Schlaich et al. (1987) for frame joints under opening moments. In order to circumvent the tensile chord reinforcement and prevent cracking of the compression chord due to radial tensile stresses, either the chord reinforcement must be extended as a loop around the corner or inclined stirrups must be adequately arranged. Other STMs are suggested in order to explain the joint behavior as illustrated in Figure 5.13b.

5.4.7 KNEE CORNER JOINTS UNDER CLOSING MOMENTS

Different STMs that were suggested by Schlaich and Schäfer (1991) are shown in Figure 5.14a. Other STMs, illustrated in Figure 5.14b, are suggested to explain the joint behavior.

5.4.8 EXTERIOR BEAM-COLUMN CONNECTIONS

The simple STM shown in Figure 5.15a suggests that the shearing and compression forces resulting from the particular load pattern are largely transmitted by a diagonal strut across the joint. In fact, there are several struts separated by diagonal cracks. It would be extremely optimistic to assume that the full compression strength could be approached in these struts. Not only are they subjected to indeterminate eccentricities, but they are also exposed to transverse tensile strains. In this biaxial state of stress, a considerable reduction of compressive strength ensues (Park and Paulay, 1975). In Figure 5.15b and c, different STMs for exterior beam-column connections

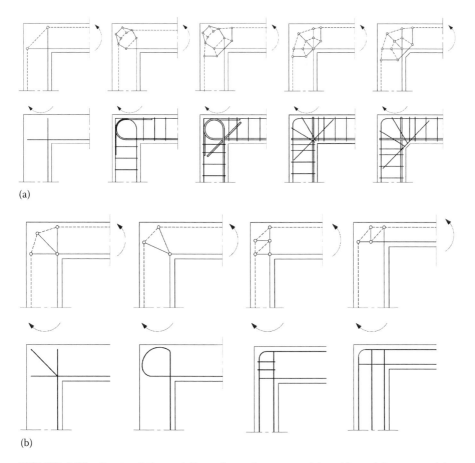

FIGURE 5.13 Strut-and-tie modeling of opening corner joint: (a) models proposed by Schlaich et al. (1987). (Adapted from Schlaich, J. et al., *J. Prestressed Concr. Inst.*, 32(3), 74, 1987.) (b) Additional suggested models.

are shown. These STMs are related to different stiffnesses of the adjacent members connected with the joint.

5.4.9 Tee Beam-Column Connections

Figure 5.16a shows the appropriate STM for the case of gravity load. In a laterally loaded frame, the forces acting on a T-joint can be idealized as shown by the simple STM in Figure 5.16b (or Figure 5.16c). This model suggests that the internal forces resulting from the particular load pattern are largely transmitted by a diagonal strut across the joint.

Additional STMs of tee beam-column connections for different geometry and detailing of the adjacent members connected with the joint are shown in Figure 5.16d and e.

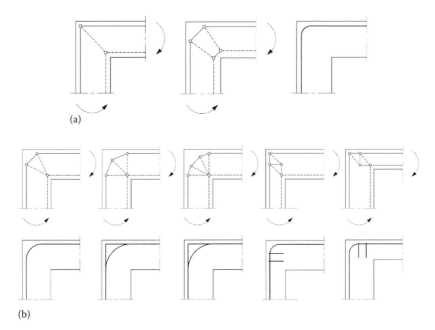

FIGURE 5.14 Strut-and-tie modeling of a closing corner joint: (a) models proposed by Schlaich and Schäfer (1991). (Adapted from Schlaich, J. and Schäfer, K., *J. Struct. Eng.*, 69(6), 113, 1991.) (b) Additional suggested models.

5.4.10 INTERIOR BEAM-COLUMN CONNECTIONS

The appropriate STM for the case of gravity load is shown in Figure 5.17a. For a laterally loaded connection, the simple STM shown in Figure 5.17b suggests that the shearing and compression forces resulting from the particular load pattern are largely transmitted by a diagonal strut across the joint. The STM shown in Figure 5.17c considers both the strut and truss mechanisms' contribution in transferring shear.

5.5 AN ILLUSTRATIVE DESIGN EXAMPLE

In this section, the design procedure of a beam with an abrupt change in thickness, Figure 5.18, is illustrated. The width of the beam $b = 300\,\text{mm}$ and the other dimensions (in mm) are shown in the figure. The factored design value of the applied force is $F_u = 1200\,\text{kN}$. The concrete cylinder strength is $f_c' = 30\,\text{MPa}$ and the steel yield stress is $\sigma_o = 460\,\text{MPa}$. The failure criteria adopted by the ACI 318-08 (2008), as illustrated in Chapter 3, is adopted in the design procedure of this example. The strength reduction factor for the effective concrete of struts and nodes and for ties' yield stress is 0.75. The design procedure is given in the following steps.

5.5.1 REACTIONS AND STRAINING ACTIONS

$$R_A = R_B = 1200\,\text{kN}$$

The bending moment and shearing force diagrams are shown in Figure 5.18.

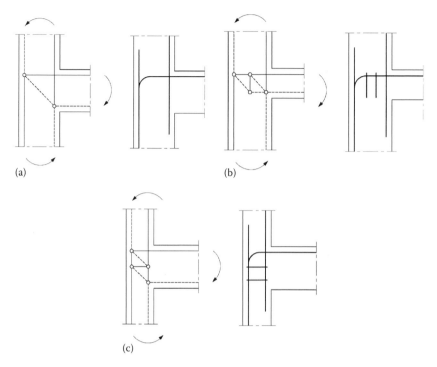

FIGURE 5.15 Suggested STMs and corresponding detailings for exterior beam-column connections with beam to column thickness: (a) between 0.67 to 1.5, (b) less than 0.67, and (c) greater than 1.5.

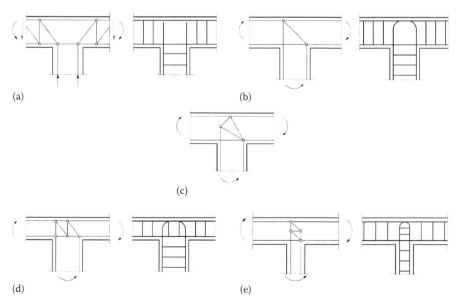

FIGURE 5.16 Suggested STMs and corresponding detailings for tee beam-column connections: (a) due to gravity loads and (b) through (e) due to lateral loads.

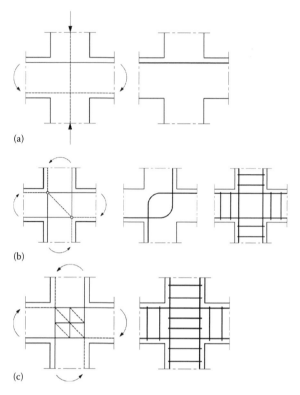

FIGURE 5.17 Suggested STMs and corresponding detailings for interior beam-column connections: (a) due to gravity loads and (b) and (c) due to lateral loads.

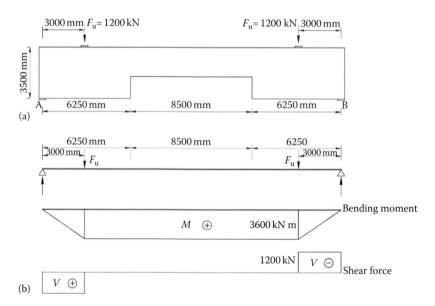

FIGURE 5.18 Transfer girder example: (a) dimensions and (b) straining actions.

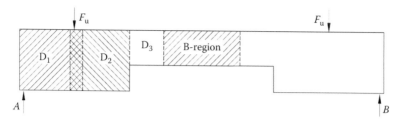

FIGURE 5.19   D- and B-regions of the transfer girder example.

5.5.2   D- AND B-REGIONS

Based on St. Venant's principle the D-regions are determined. Figure 5.19 illustrates both D- and B-regions.

5.5.3   DIMENSIONING OF B-REGION

The bending moment of the B-region is a constant value of $M_u = 3600\,\text{kN m}$ and the shear is zero. If the contribution of compression reinforcement is neglected (Figure 5.20),

$$M_u = 3600\,\text{kN m} = \phi M_n = \phi\left[0.85 f_c' ab\left(d - \frac{a}{2}\right)\right]$$

Since the moment is a constant value, the dividing section between the B- and D-regions requires the use of the strength reduction factor of the STM, that is, $\phi = 0.75$ instead of 0.9 (the value used in flexure). When substituting $b = 300\,\text{mm}$, $d \approx 1900\,\text{mm}$, and $f_c' = 30\,\text{MPa}$, the value of a obtained is 365 mm. The lever arm is $Y_{CT} = d - a/2 = 1718\,\text{mm}$. From moment equilibrium, Figure 5.20,

$$C_1 = T_1 = \frac{M_u}{Y_{CT}} = \frac{3600 \times 10^6}{1718} = 2095\,\text{kN}$$

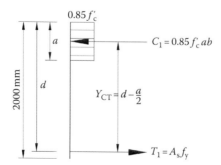

FIGURE 5.20   Force equilibrium of B-region.

$$A_{s1} = \frac{T_1}{\phi f_y} = \frac{2095 \times 10^3}{0.75 \times 460} = 6073 \, \text{mm}^2 = 16\text{D}22$$

5.5.4 ESTABLISH AN STM

The load path and the corresponding STM are shown in Figure 5.21. The compression block at the dividing section between the B- and D-regions is subdivided into two parts: one to balance the force C_3 and the other to balance the horizontal component of the force C_4. From equilibrium of the D-region, Figure 5.21, and with the aid of the geometric relations in Figure 5.22,

$$C_3 = 1200 \times \frac{3.0}{3.3} = 1091 \, \text{kN}$$

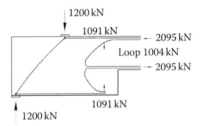

Load paths and factored loads

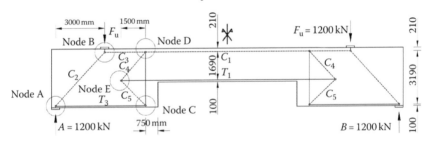

FIGURE 5.21 Load path and STM of the transfer girder.

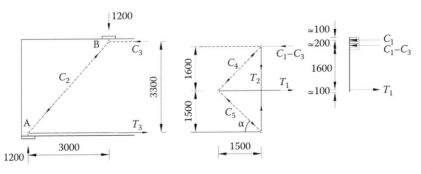

FIGURE 5.22 Geometric relations of the STM.

$$T_3 = C_3 = 1091\,\text{kN}$$

$$C_2 = \sqrt{(1200)^2 + (1091)^2} = 1622\,\text{kN}$$

If the angle α is assumed equal to $45°$, then, the force T_2 will be

$$T_2 = T_3 = 1091\,\text{kN}$$

which leads to

$$C_5 = \sqrt{T_2^2 + T_3^2} = 1543\,\text{kN}$$

$$C_4 = \sqrt{(C_1 - C_3)^2 + T_2^2} = 1483\,\text{kN}$$

5.5.5 EFFECTIVE CONCRETE STRENGTH FOR THE STRUTS

$$f_{cu} = 0.85\beta_s f_c'$$

Struts C_1 and C_3 are prismatic stress fields for which $\beta_s = 1.0$; hence, $f_{cu} = 25.50\,\text{MPa}$

Struts C_2, C_4, and C_5 are bottle-shaped stress fields for which $\beta_s = 0.75*$; hence, $f_{cu} = 19.13\,\text{MPa}$

5.5.6 EFFECTIVE CONCRETE STRENGTH FOR THE NODES

$$f_{cu} = 0.85\beta_n f_c'$$

Node A is a C-C-T node; therefore, $\beta_n = 0.8$; hence, $f_{cu} = 20.4\,\text{MPa}$
Node B is a C-C-C node; therefore, $\beta_n = 1.0$; hence, $f_{cu} = 25.5\,\text{MPa}$
Node C is a C-T-T node; therefore, $\beta_n = 0.6$; hence, $f_{cu} = 15.3\,\text{MPa}$
Node D is a C-C-T node; therefore, $\beta_n = 0.8$; hence, $f_{cu} = 20.4\,\text{MPa}$
Node E is a C-C-T node; therefore, $\beta_n = 0.8$; hence, $f_{cu} = 20.4\,\text{MPa}$

5.5.7 NODE A

The bearing plate length, $a* = R_A / \phi f_{cu} b = 1200 \times 10^3 / 0.75 \times 20.4 \times 300 = 261$ mm. Use a bearing plate of the dimension 300×300 mm. With reference to Figure 5.23, if two layers of reinforcement are used, then, $u = 150$ mm, and from Figure 5.22 $\tan\theta_A = (3.3/3.0)$, which gives $\theta_A = 47.7°$. The width of strut C_2 at node A, Figure 5.23, is

$$w_{C_2} = a* \sin\theta_A + u\cos\theta_A = 323\,\text{mm}$$

---

* Transverse reinforcement to resist the lateral tension will be provided.

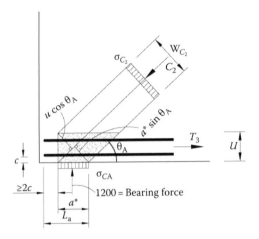

FIGURE 5.23   Node A.

The stress in the strut is

$$\sigma_{C_2} = \frac{C_2}{w_{C_2}b} = \frac{1622}{323 \times 300} = 16.74 \, \text{MPa} > \phi f_{cu}(= 15.3 \, \text{MPa})$$

Either increase u or use a larger bearing plate. Upon using a bearing plate $400 \times 300 \, \text{mm}$ and redoing the calculations, it is found that $w_{C_2} = 397 \, \text{mm}$ and $\sigma_{C_2} = 13.62 \, \text{MPa} < \phi f_{cu}$; the bearing plate size is adequate.

5.5.8   NODE B

Try a bearing plate with dimension $300 \times 300 \, \text{mm}$ (Figure 5.24). The bearing stress is $1200 \times 10^3/300 \times 300 = 13.33 \, \text{MPa}$, which is less than $\phi f_{cu}(= 19.13)$. The depth of the compression block of strut C_3 is

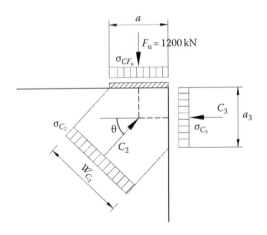

FIGURE 5.24   Node B.

$$w_{C_3} = \frac{C_3}{\phi f_{cu} b} = \frac{1091 \times 10^3}{0.75 \times 25.5 \times 300} = 190 \, \text{mm}$$

The width of strut C_2 will be (Figure 5.24) $(300 \sin \theta_B + 190 \cos \theta_B = 350 \, \text{mm}$, $\theta_B = \theta_A = 47.7°)$. The stress of the node at the interface of strut C_2 is

$$\sigma_{C_2} = \frac{C_2}{w_{C_2} b} = \frac{1622 \times 10^3}{350 \times 300} = 15.45 \, \text{MPa} < \phi f_{cu} (= 19.13 \, \text{MPa})$$

5.5.9 NODE C

The width of strut C_5, Figure 5.25, is

$$w_{C_5} = \frac{C_5}{\phi f_{cu} b} = \frac{1543 \times 10^3}{0.75 \times 15.3 \times 300} = 448 \, \text{mm} = x_2 \sin \alpha = x_2 \sin 45°$$

which results in $x_2 = 634 \, \text{mm}$. The reinforcement resisting T_2 should be closed stirrups and placed within a distance at least equal to x_2.

5.5.10 NODE D

The depth of the compression block of $(C_1 - C_3)$ is x_{13}, Figure 5.26, given by

$$x_{13} = \frac{C_1 - C_3}{\phi f_{cu} b} = 219 \, \text{mm}$$

The width of strut C_4, Figure 5.26, is

$$w_{C_4} = \frac{C_4}{\phi f_{cu} b} = \frac{1483 \times 10^3}{0.75 \times 20.4 \times 300} = 323 \, \text{mm} = x_{22} \sin \beta + x_{13} \cos \beta$$

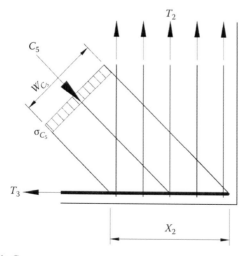

FIGURE 5.25 Node C.

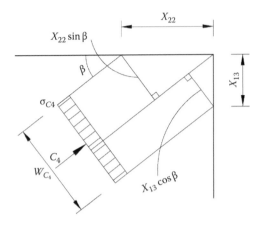

FIGURE 5.26   Node D.

From geometry, Figure 5.22, the angle $\beta = \tan^{-1}(160/150) = 46.85°$, which gives $x_{22} = 237\,\text{mm}$. x_{22} is the minimum distance within which the closed stirrups resisting T_2 should be placed. Since $x_{22} < x_2$, the value of x_2 should be used, which increases the width of strut C_4 to $612\,\text{mm}$.

5.5.11   NODE E

The widths of struts C_4 and C_5 at the node, Figure 5.27, can be calculated in the same manner as before based on the design strength of the node. However, the design strength of strut C_5, $f_{cu} = 19.13\,\text{MPa}$, is less than the design strength of the node,

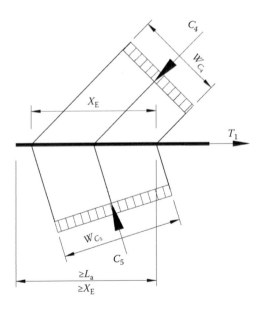

FIGURE 5.27   Node E.

$f_{cu} = 20.4$ MPa; therefore, for strut safety the lower value should be used in dimensioning the node. The width of strut C_5 should therefore be $C_5/(\phi f_{cu} b) = 359$ mm, which gives a projection in the direction of T_1, $x_E = 507$ mm.

For checking the node with regard to C_4, the width of the strut is $x_E \sin \beta = 370$ mm, and the corresponding stress is $C_4/(370 \times 300) = 13.36$ MPa $< \phi f_{cu}$ (= $0.75 \times 20.4 = 15.3$ MPa). The anchorage of the reinforcement resisting T_1 should be extended to the largest of anchorage length or x_E, from the beginning of the node (Figure 5.27).

5.5.12 Strut C_2

The width of the strut at end A is 397 mm and at end B is 350 mm; the smaller value is considered for checking the safety of the strut. The stress in the strut is therefore

$$\frac{1622 \times 10^3}{350 \times 300} = 15.45 \, \text{MPa} > \phi f_{cu} (0.75 \times 19.13 = 14.35 \, \text{MPa})$$

The width of the strut at end B should be increased to 377 mm, which can be achieved by increasing the length of the bearing plate to 337 mm. Therefore, the length of the bearing plate at node B, Figure 5.24, should be increased to 350 mm instead of 300 mm.

The transverse reinforcement of the strut is required to resist a total force T_{C_2}. From the STM of Figure 5.28,

$$T_{C_1} \approx \left(\frac{1}{2} \times \frac{C_1}{2}\right) \times 2 = \frac{C_1}{2} = \frac{1622}{2} = 811 \, \text{kN}$$

Thus, the total reinforcement required perpendicular to the strut is $811 \times 10^3 / (0.75 \times 460) = 2351$ mm^2. The length of the strut is 4460 mm; hence, the required transverse reinforcement perpendicular to the strut is 0.527 mm^2/mm. This can be covered with a skin reinforcement of vertical bars of diameter 12 mm every 200 mm and horizontal bars of diameter 12 mm every 200 mm, on each side, in addition to interior open stirrups of diameter 10 mm every 400 mm. The larger reinforcement

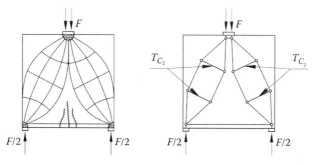

FIGURE 5.28 Stress deviation in a bottle-shaped stress field.

is assigned to the vertical direction since it is more effective in substituting for the inclined reinforcement because the angle θ_A is less than 45°. With reference to Figure 3.50, the used transverse steel is equivalent to inclined reinforcement $\Sigma A_{si} \sin \gamma_i/bs_i = 0.0053 > 0.003$ (the ACI minimum value for $f'_c \leq 44$ MPa).

5.5.13   STRUT C_3

Since this strut is a prismatic stress field, it is adequate to check the nodes only.

5.5.14   STRUT C_3

The width of the strut at nodes D and E is 612 and 370 mm, respectively. The smaller width is used in checking the strut, which gives a stress

$$\frac{C_4}{370 \times 300} = 13.36 \text{ MPa} < \phi f_{cu}(0.75 \times 19.13 = 14.35 \text{ MPa})$$

The transverse reinforcement of the strut is required to resist a total force $C_4/2 = 741.5$ kN, which requires a total reinforcement equal to 2149 mm², perpendicular to the strut. The length of the strut is 2193 mm; hence, the required transverse reinforcement is 0.98 mm²/mm, perpendicular to the strut. This is covered with the predetermined skin reinforcement of vertical bars of diameter 12 mm every 200 mm and horizontal bars of diameter 12 mm every 200 mm, on each side. Also, the used transverse steel is equivalent to inclined reinforcement $\Sigma A_{si} \sin \gamma_i/bs_i = 0.00532 > 0.003$ (the ACI minimum value for $f'_c \leq 44$ MPa).

5.5.15   STRUT C_5

The width of the strut at nodes C and E is 448 and 359 mm, respectively. The smaller width is used in checking the strut, which gives a stress

$$\frac{C_5}{359 \times 300} = 14.33 \text{ MPa} < \phi f_{cu}(0.75 \times 19.13 = 14.35 \text{ MPa})$$

The transverse reinforcement of the strut is required to resist a total force $C_4/2 = 771.5$ kN, which requires a total reinforcement equal to 2236 mm², perpendicular to the strut. The length of the strut is 2121 mm; hence, the required transverse reinforcement is 1.05 mm²/mm, perpendicular to the strut. This is covered with the predetermined skin reinforcement of vertical bars of diameter 12 mm every 200 mm and horizontal bars of diameter 12 mm every 200 mm, on each side. Also, the used transverse steel is equivalent to inclined reinforcement $\Sigma A_{si} \sin \gamma_i/bs_i = 0.00533 > 0.003$ (the ACI minimum value for $f'_c \leq 44$ MPa).

5.5.16   CHECKING THE STRENGTH OF THE B-REGION AND TIE T_1

$$\phi M_n = \left[C_3 \left(d - \frac{190}{2} \right) + (C_1 - C_3) \left(d - 190 - \frac{219}{2} \right) \right] = 3576 \text{ kN m} < M_u (= 3600 \text{ kN m})$$

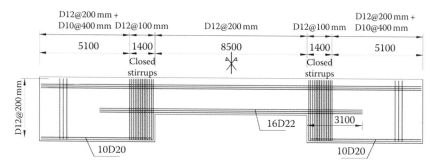

FIGURE 5.29 Reinforcement layout.

The slight difference between M_u and ϕM_n can be covered by additional compression and tension reinforcement of an area 38 mm² on each side. Still, the value of 16D22 is adequate as a tension reinforcement of T_1.

5.5.17 Tie T_2

The reinforcement required to resist this force is $T_2/(0.75 \times 460) = 3162$ mm², which can be covered with two-branches closed stirrups 14D12.

5.5.18 Tie T_3

The reinforcement required to resist this force is $T_3/(0.75 \times 460) = 3162$ mm², which can be covered with 10D20.

5.5.19 Reinforcement

The used skin reinforcement D12 at 200 mm on both sides in the vertical and horizontal directions gives a ratio of 0.38% both vertically and horizontally, which is adequate. The final beam reinforcement is illustrated in Figure 5.29.

5.6 HISTORICAL SKETCH

With the extension of the limit design theorems to continuous media by Drucker, Greenberg, and Prager in 1952, applications of the powerful limit analysis techniques were expanded to plates and shells for both metal and reinforced concrete materials as well as soil mechanics. The yield line theory for flexure analysis of reinforced concrete slabs is the most successful application of upper-bound method of perfect plasticity to structures. Also, for flexure analysis of reinforced concrete beams and frames, the limit analysis has become standard since the 1950s. Considering shear problems, however, very few theoretical advances had been made before the 1970s. The application of the theory of plasticity to the design of members under shear and torsion began in the 1970s, especially by Thürlimann et al. (1975), Thürlimann et al. (1983), Nielsen et al. (1978), and Nielsen (1984). This also formed the basis for the method of STMs after the work of Schlaich et al. (1987) and Schlaich and Schäfer (2001), which formed the basis for the contents of this chapter.

The development of the STM method has brought a major breakthrough in designing a consistent theory in the design concept covering both the D-regions and the B-regions with similar models. The method provides a formal design procedure for detailing in design in particular. All these developments were brought out in the state-of-the-art report on shear by the ASCE-ACI Committee 445 in 1998. The ACI Committee 318 introduced the method of STMs into its 2002 ACI code. Appendix A of the ACI-318-2002 documented this international development in research, which formed the basis for other codes around the world. This step is an important milestone in the direction toward the development of a more consistent design concept. It triggered the acceptance and wide use of STMs for daily use.

The concept of using the method of STMs to inelastic-reinforced concrete analysis was introduced and illustrated for the first time in 1961 by Drucker in his estimate of the load-carrying capacity of a simply support ideal reinforced concrete beam. It took a great physical insight to fully understand the fundamental difference between a tension-weak material like concrete or soils in its load-carrying capacity through arching compared to that of a ductile material like metal through flexure or bending. Thanks to this revolutionary thinking, the concept of STM was born. The subsequent development, refinement, and expansion resulted in the modern techniques of STMs for detailing design of shear, torsion, joints, and bearing in structural concrete in a consistent manner.

In the following, we shall present Drucker's original *simple beam model* (1961) to illustrate his concept of lower- and upper-bound techniques of limit analysis as applied to a reinforced concrete beam in Figure 5.30. For simplicity, he assumed the concrete beam to have negligible weight and zero-tensile strength, so that it acts as a very flat arch. The outward thrust of the arch is shown in Figure 5.30a as being taken

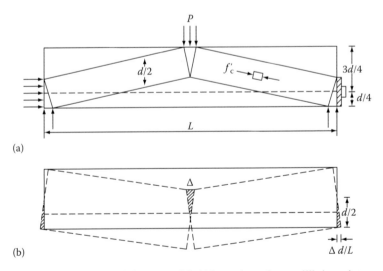

(a)

(b)

FIGURE 5.30 Drucker's simple beam model: (a) lower-bound or equilibrium picture of arch action and (b) kinematical picture of collapse mechanism. (Modified from Drucker, D.C., *Publ. IABSE*, 21, 45, 1961.)

by a steel tension tie between two end plates bearing on the concrete. The steel was assumed unbonded. Efficient use of material would seem to dictate at first that, at the ultimate or collapse load both steel and concrete should be at their yield and failure stresses, respectively.

Since the equilibrium distribution of stress in the concrete and the steel as shown in Figure 5.30a was nowhere tensile in the concrete and was everywhere at or below yield, the beam would not collapse at this load or would be just at the point of collapse, according to the lower-bound theorem. This approach so far focused on the lower-bound equilibrium technique and thus it might underestimate the strength of the beam.

Figure 5.30b shows a kinematical picture associated with an assumed plastic failure mechanism, which gave an upper bound on the collapse load. The assumed failure mechanism as drawn showed the stretching or yielding of the steel tie and the crushing plastically of the shaded areas of concrete at the ends as well as at the center. This failure mechanism resulted in an upper-bound solution, which turned out to be equal to the lower-bound solution of Figure 5.30a. Thus, Drucker obtained the correct answer for the idealized beam according to the limit theorem despite the fact that either the stress field as constructed in Figure 5.30a or the plastic collapse mechanism as assumed in Figure 5.30b was the real stress distribution or the real failure mechanism, respectively. This simple example clearly illustrated Drucker's basic concept and power of limit analysis as applied to reinforced concrete structures. It also physically showed how the load was carried in a composite structure through arching for tension-weak concrete and stretching for tension-strong steel to its supports or foundations.

Professor J. Schlaich and his colleagues and students at the Institute of Structural Design, University of Stuttgart, had worked for decades on the application of the method of STM for a uniform treatment to all D-regions in order to achieve a consistent treatment of B- and D-regions. Their contributions have been greatly influential, particularly in exploring and identifying all different D-regions based on geometry and boundary conditions for a uniform treatment of these regions. In addition, they illustrated, based on logic and transparency or from mechanics, how D-regions can be modeled. Besides, they set simple, but reliable, failure criteria of STM elements. The paper by Schlaich, Schäfer, and Jennewein is a landmark in this area (Schlaich et al., 1987). The efforts of Professor J. Schlaich and his group from the University of Stuttgart have made the method of STM widely recognized in the academia and widely adopted in practice.

REFERENCES

ACI 318-08, 2008, *Building Code Requirements for Structural Concrete and Commentary*, American Concrete Institute, Farmington Hills, MI.

ASCE-ACI 445, 1998, Recent approaches to shear design of structural concrete. State-of-the-Art-Report by ASCE-ACI Committee 445 on Shear and Torsion. *ASCE Journal of Structural Engineering*, 124, 12, 1375–1417.

Drucker, D. C., 1961, On structural concrete and the theorems of limit analysis, *Publications IABSE*, 21, 45–49.

Egyptian Code for the Design and Construction of Concrete Structure, ECP 203-2006, Ministry of Housing, Utilities and Urban Communities, National Housing and Building Research Center, Cairo, 2006.

MacGregor, J. G., 1984, *Challenges and Changes in the Design of Concrete Structures*, Concrete International, ACI, Detroit, MI.

Nielsen, M. P., 1984, *Limit Analysis and Concrete Plasticity*, Prentice Hall, Englewood Cliffs, NJ, 420pp.

Nielsen, M. P., Braestrup, M.W., Jensen, B. C., and Bach, F., 1978, *Concrete Plasticity: Beam Shear-Shear in Joints-Punching Shear*, Special Publication, Danish Society for Structural Science and Engineering, Technical University of Denmark, Lyngby, Denmark, p. 1–129.

Park, R. and Paulay, T., 1975, *Reinforced Concrete Structures*, John Wiley & Sons, New York.

Schäfer K. and El-Metwally, S. E., 1994, On the role of discontinuity regions detailing in the safety of concrete structures, *Proceedings of the Fifth International Colloquium on Concrete in Developing Countries*, Cairo, Egypt, January 2–6, pp. 43–55.

Schlaich, J. and Schäfer, K., 1991, Design and detailing of structural concrete using strut-and-tie models, *Journal of the Structural Engineer*, 69, 6, 113–125.

Schlaich, J. and Schäfer K., 1993, The design of structural concrete, *IABSE Workshop*, New Delhi, India.

Schlaich, J. and Schäfer, K., 2001, Konstruieren in Stahlbetonbau (Detailing of Reinforced Concrete), *Betonkalender* 90, Tiel II, pp. 311–492, Ernst & Sohn Verlag, Berlin, Germany.

Schlaich, J., Schäfer, K., and Jennewein, M., 1987, Toward a consistent design of structural concrete, *Journal of the Prestressed Concrete Institute*, 32, 3, 74–150.

Thürlimann, B., Grob, J., and Lüchinger, P., 1975, Torsion, Biegung und Schub in Stahlbetonträgern. *Fortbildungskurs für Bauingenieure*, Institut für Baustatik und Konstruktion, ETH Zurich, Switzerland.

Thürlimann, B., Marti, P., Pralong, J., Ritz, P., and Zimmerli, B., 1983, Anwendung der Plastizitatstheorie auf Stahlbeton. *Fortbildungskurs für Bauingenieure*, Institut für Baustatik und Konstruktion, ETH Zurich, Switzerland.

6 Toward Advanced Analysis for Steel Frame Design

6.1 THE ROLE OF THE EFFECTIVE LENGTH FACTOR K IN DESIGN

6.1.1 GENERAL

The purpose of structural design is to produce a physical structure capable of withstanding the environmental conditions to which it may be subjected. Many factors affect the design process, from loading to foundation to dimensions of layout to risk and cost, but basically the ultimate design is a reflection of the properties of the structural material and the geometrical imperfection of its structural members and in particular its mechanical properties and its residual stresses induced in the structural members during the manufacture and fabrication processes, which define the characteristic response of the material and the member of the force field of its environment.

In the present engineering design practice, there is a fundamental two-stage process in the design operation: first, the forces acting on each structural member in the structure must be calculated and second, the load-carrying capacity of each of these structural members to those forces acting on it must be determined. The first stage involves an analysis of the distribution of forces and moments acting on each of these structural members and the second stage involves knowledge of the load-carrying capacity of these members to resist these forces and moments acting on them. The more comprehensive this knowledge is, the more exact will be the design and the more reliable will be the structure.

Since the load-carrying capacity of structural members depends on the type of load acting on the member, geometrical imperfections, properties of material, and residual stresses, the knowledge of load-carrying capacity of these structural members has been determined mostly on the basis of full-scale tests in the form of pin-ended column strength curves for axially loaded members, simply supported beam strength curves for bending dominated members, and beam-column interaction curves for members under combined axial force and bending moment. These member strength curves are formally coded as the member strength curves or equations for design practice.

Having divided the structural members in a framed structure into three classes, namely, columns, beams, and beam-columns, and determined their respective strengths by full-scale tests with ideal end or boundary conditions, the next stage must be to drastically simplify the material behavior under stresses in such a way as to readily assist the professional engineer to analyze the stress distribution in the structure in order to size up the structural members in a framed structure. At present, the practicing engineer bases design primarily on the simple model of linear elasticity of the material for those early designs based on the allowable stress method.

In this process, time-dependent effects of material are assumed insignificant. This is indeed a drastic simplification of material properties over long-term behavior. Therefore, with this time-independent simplification, the design process is now focused on reducing the level of stresses or the structure is loaded only in the working load level. A large safety factor is therefore used to adjust the design to take account of inelastic features in the material to avoid failure. Linear elastic analysis has been predominantly used in engineering practice. First-order linear elastic analysis has been the hallmark of structural engineering in early years, while the second-order linear elastic analysis for a structural system has been developed and increasingly utilized in recent years.

6.1.2 Elastic Structural Analysis and K Factor

The boundary conditions of a *framed member* in a frame structure are quite different from those of an isolated member that are used as the basis for the development of column strength curves (pin–pin end conditions) or beam strength curves (simply supported end conditions). In order to size up frame members, the members' boundary conditions must be adjusted to the equivalent pin–pin end conditions for the case of column design, for example, so that the column strength curves can be properly used in determining the required size of the framed member under consideration (Figure 6.1).

To achieve this equivalency, the *effective length factor* or the K factor has been widely used in the past to relate the pin-ended column strength curves to the framed member design in a structural system. The effective length method is a good method for the design of framed structures. This method has been widely used for the development of modern steel design codes, including the allowable stress design (ASD) and the plastic design (PD) in early years and the load-and-resistance-factor design in more recent years (Salmon and Johnson, 1990).

The critical load of any beam-column of any boundary conditions, P_{cr}, can be written in a unified form as follows:

$$P_{cr} = \frac{\pi^2 EI}{L_b^2} \tag{6.1}$$

where L_b is the buckling length of the column, which is the distance between the two inflection points of the beam-column in its deformed shape. The critical load of a two-hinged-ends column is the Euler load, P_E, given by

$$P_E = \frac{\pi^2 EI}{L^2} \tag{6.2}$$

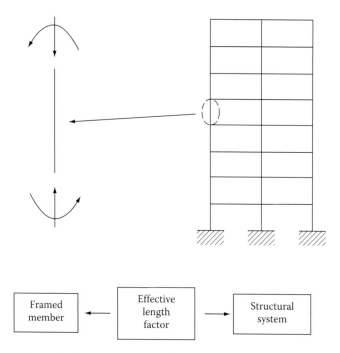

FIGURE 6.1 Interaction between a structural system and its component member.

Thus, the critical load, P_{cr}, can be related to the Euler load, P_E, as follows:

$$P_{cr} = \frac{P_E}{(L_b/L)} = \frac{P_E}{K^2}$$ (6.3)

where the factor K is the ratio between the column buckling length, L_b, and its original length, L. The K factor values for standard cases of boundary conditions are given in Table 2.2, and for a column in a framed system, it can be assessed with the aid of alignment charts based on the bracing condition of the structural system (Figure 2.21).

The adoption of elastic structural analysis with K factor for steel design may be divided into two stages of progress. The simpler first stage of progress with K factor in the design process is the first-order elastic analysis with amplification factors to include the second-order effects as generally provided by the specifications (LRFD, 1986). The next logical stage of progress is a direct second-order elastic analysis without the use of amplification factors for second-order effects (AISC-LRFD, 2005). Both methods are based on the formation of the first plastic hinge defined as the failure of the system.

6.1.3 DESIGN WITH K FACTOR

In the *allowable stress design method,* a first-order elastic analysis is performed to obtain the member forces. The second-order effects, the P–Δ effect and the P–δ

effect, are considered in the design stage through the amplification factors. In order to carry out member design, the influence of the structural system on member capacity has to be introduced with the aid of the effective length factor K (Figure 6.1). The value of K factor for the sizing of different members can be obtained with the aid of alignment charts.

For the *plastic design method*, a first-order elastic–plastic analysis is performed in order to obtain the internal forces. In this analysis, inelasticity is accounted for through the simple plastic-hinge concept. Hence, inelastic force redistribution throughout the structural system is achieved; however, the spread of yielding is not considered. In addition, stability effects and geometric nonlinearity, are not accounted for in this analysis and are implemented in the member design through the amplification factors. This means that this analysis is superior only to the linear elastic analysis in the force redistribution. In this design method, the K factor still has to be implemented in the design of members. The value of K for the different members can be determined with the aid of alignment charts or any other method.

In the *LRFD design method*, the designer has the option to carry out a first-order elastic analysis and use the amplification factors to account for the stability effects, or carry out a second-order elastic analysis in which the stability effects are implemented. The ultimate strength of beam-column members considering gradual yielding is implicit in the design interaction equations. For member design, the effective length factor K has to be determined with the aid of alignment charts.

6.1.4 LIMITATIONS

Despite the popularity of the K factor method in the past decades, the approach has several major limitations and shortcomings.

The first of these limitations is that the K factor method does not give an accurate indication of the safety factor against failure because it does not consider the interaction of strength and stability between the member and the structural system in a direct manner. It is a well-recognized fact that the actual failure mode of the structural system often does not have any resemblance whatsoever to the elastic buckling mode of the structural system, which is the basis for the determination of the effective length factor K.

The second and perhaps the most serious limitation is probably the rationale of the current two-stage process in design: elastic analysis is used for the determination of distribution of forces acting on each member of a structural system, whereas the member ultimate strength curves are developed for design either on the basis of full-scale tests or by inelastic analysis with each member being treated as an isolated component (Chen and Atsuta, 2007a,b). There is no verification of the compatibility between the isolated member and the member as part of a frame. The individual member strength equations as specified in specifications are not concerned with system compatibility. As a result, there is no explicit guarantee that all members will sustain their design loads under the geometric configuration imposed by the framework.

The other limitations of the effective length method include the difficulty in computing a K factor, which is not user friendly for computer-based design, and the inability of the method to predict the actual strength of a framed member, among others. To this end, there is an increasing tendency of the need for practical analysis/ design methods that can account for the compatibility between the member and the system. With the rapid development of computing power and the availability of desktop computers and user-friendly software, the development of an alternative method to a direct design of structural system without the use of K factors becomes more attractive and realistic.

As mentioned previously, the effective length factor will generally yield good designs for framed structures, but it does have the following drawbacks:

1. It cannot capture accurately the interaction between the structural members and the structural system behavior and strength.
2. It cannot reflect the proper inelastic redistributions of internal forces in a structural system.
3. It cannot predict the failure modes of a structural system.
4. It is not easy to be implemented in an integrated computer design application with the use of alignment charts in the K factor calculation process.
5. It is generally a time-consuming process requiring separate member capacity checks with different K factors for different framed members.

Even for the most recent AISC-LRFD procedures, similar difficulties are encountered when performing a seismic design because the checks of the same strength interaction equations must be performed (AISC-LRFD, 2005). Some of the difficulties are even more so on the seismic designs since additional questions such as the following are raised:

1. How is the structure going to behave during the earthquake?
2. Which part of the structure is the most critical area?
3. What will happen if part of the structure yields or fails?
4. What might happen if forces greater than the code has specified occur?

None of these questions can be answered by the conventional load-resistance-factor design (LRFD) method with K factors. With the rapid advancement of computing power, the second-order inelastic analysis approach or the so-called *advanced analysis* provides an alternative approach to structural analysis and design without the K factor. Nevertheless, this approach consumes tremendous computation efforts, and in order to overcome such a demand, the practical advanced analysis has been developed. The practical advanced analysis method presented in this chapter is an elastic-plastic-hinge-based analysis, modified to include the effects of geometry imperfections, spread of plasticity, residual stresses, and semi-rigid connections (Chen, 2009). In this approach, all these aforementioned drawbacks of design using the effective length factor K are overcome and there is no need to compute this factor.

6.2   METHODS OF ADVANCED ANALYSIS

6.2.1   DEFINITIONS

Advanced analysis refers to any method that captures the strength and stability of a structural system and its individual members in such a way that separate member capacity checks are not required. Further, advanced analysis accounts for (1) material inelasticity, (2) stability effects, (3) residual stresses, (4) geometric imperfections, (5) local buckling, (6) connection behavior, (7) end restraint, and (8) interaction with the foundations. Usually, these analyses are also referred to more formally as the *second-order inelastic analysis for frame design*. In this direct approach, there is no need to compute the effective length factor K since separate member capacity checks encompassed by the specification equations are not required. This design approach is illustrated in Figure 6.2 and marked as the direct analysis and design method.

With the present computing power, it is a rather straightforward process to combine the theory of stability (Chen and Lui, 1986) with the theory of plasticity (Chen and Han, 2007) for a structural system analysis (Chen and Toma, 1994). The real challenge is therefore to make this type of new approach to design work and compete with the current methods in engineering practice (Chen and Lui, 1991, 2005). The main distinction between advanced analysis and conventional methods is that advanced analysis can predict the structural system strength, whereas others predict the member strengths. Advanced analysis uses a format associated with the structural system rather than individual members.

Since advanced analysis provides the most benefit in modern frame design, specification provisions of this type are required for practical use. The provisions for advanced analysis in some codes, such as the Eurocode and Australian limit state specifications, postulate that if all the significant planar behavioral effects are modeled properly in the analysis, the checking of conventional beam-column equations is not required. Advanced analysis uses the same resistance factors and load

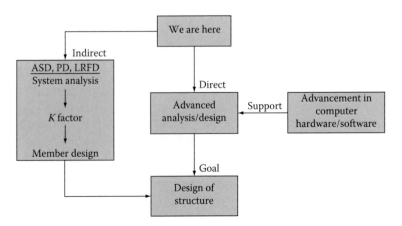

FIGURE 6.2   Indirect and direct analyses and design methods.

factors as those of the LRFD method. In this manner, a uniform reliability can be achieved by reflecting the degree of uncertainty of different loads and combinations of loads.

6.2.2 Elastic–Plastic-Hinge Method

The *elastic–plastic-hinge method* is the simplest approximation of inelastic behavior of a material by assuming all its inelastic effects concentrated at the plastic-hinge locations. In this idealization, it is assumed that the element remains elastic except at the ends where zero-length plastic hinges can form. This method accounts for inelasticity but not the spread of yielding or plasticity at sections, or the influence of residual stresses.

Here, as discussed in previous sections for elastic analysis, the elastic–plastic hinge method may be divided into first-order and second-order plastic analyses depending on the geometry used to form the equilibrium equations. For the first-order elastic-plastic-hinge analysis, the undeformed geometry is used, and nonlinear geometric effects are neglected. As a result, the predicted ultimate load is the same as conventional rigid-plastic analysis. In the second-order elastic–plastic analysis, the deformed shape is considered and geometric nonlinearities can be included using stability functions, which enables the use of only one beam-column element per member to capture the second-order effects. A comprehensive presentation of the plastic design and second-order analysis method can be found in Chen and Sohal (1995).

The second-order elastic–plastic-hinge analysis is only an approximate method (Liew, 1992; White, 1993). For slender members whose dominant failure mode is elastic instability, the method provides a good approximation; but for stocky members as well as for beam-column elements subjected to combined axial load and bending moment, this method overestimates the actual strength and stiffness in the inelastic range due to spread of yielding effects. This method is therefore a good first approximation of the second-order inelastic analysis for frame design within the applicable range described above. It requires further refinement before it can be recommended for analysis of a wide range of framed structures (White and Chen, 1993).

6.2.3 Refined Plastic-Hinge Method

The *refined plastic-hinge methods* are based on some simple modifications of the elastic–plastic-hinge method described in the preceding section. The notional load concept is first introduced in the conventional elastic–plastic-hinge method by applying additional fictitious equivalent lateral loads to account for the influence of residual stresses, member imperfections, and distributed plasticity that are not included in the conventional procedures. With certain modifications, this refined approach is accepted by the European Convention for Constructional Steelwork (ECCS, 1991), the Canadian Standard Association (1989, 1994), and the Australian Standard (1990). However, Liew's research (1992) shows that this method under-predicts the strength in the various leaning column frames by more than 20% and overpredicts the strength up to 10% in the isolated beam-columns subject to axial forces and bending moments.

6.2.4 PLASTIC-ZONE METHOD

The *plastic-zone method* is considered to be the "exact" method since it is based on the most refined finite element (FE) analysis for a structural system. This plastic-zone analysis has been well developed and documented by a research team at Cornell University over two decades under the leadership of Professor McGuire (1992). The team members included R. D. Ziemian (1992), D. W. White (1993), M. N. Attala, and G. G. Deierlein (1994), among others. As a comparison, the elastic–plastic-hinge model is considered to be the simplest, while the elastic–plastic-zone model exhibits the greatest refinement. The exact solutions were verified with some benchmark beam-column tests. This method is ideal for the verification of other simplified methods developed for engineering practice.

In the plastic-zone method, frame members are discretized into FEs, and the cross section of each FE is subdivided into many fibers as shown in Figure 6.3. The deflection at each division point along a member is obtained by numerical integration. The incremental load-deflection response at each loading step, which updates the geometry, captures the second-order effects. The residual stress in each fiber is assumed constant since the fibers are very small. The stress state at each fiber can be explicitly traced, so the gradual spread of yielding can be captured. The plastic-zone analysis eliminates the need for separate member capacity checks since it explicitly accounts for the second-order effects, spread of plasticity, and residual stresses. As a result, the plastic-zone solution is known as an "exact solution." A schematic comparison between the plastic-zone method and the other methods is given in Figure 6.4.

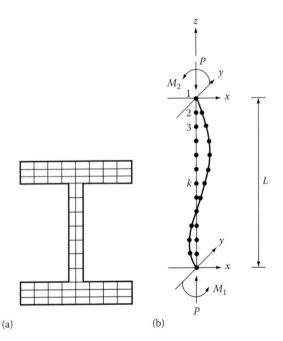

(a) (b)

FIGURE 6.3 Model of plastic-zone analysis: (a) cross-sectional discretization and (b) member division.

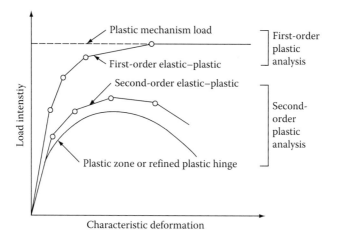

FIGURE 6.4   Load–deformation characteristics of plastic analysis methods.

There are two types of plastic-zone analyses (Chen and Kim, 1997). The first involves the use of three-dimensional finite shell elements in which the elastic constitutive matrix in the usual incremental stress–strain relations is replaced by an elastic–plastic constitutive matrix when yielding is detected. Based on the deformation theory of plasticity, the effects of combined normal and shear stresses may be accounted for. This analysis requires modeling of structures using a large number of finite three-dimensional shell elements and numerical integration for the evaluation of the elastic–plastic stiffness matrix. The three-dimensional spread-of-plasticity analysis, when combined with second-order theory, which deals with frame stability, is computationally intensive. Therefore, it is best suited for analyzing small-scale structures and providing the detailed solutions for member local instability and yielding behavior if required. Since a detailed analysis of local effects in realistic building frames is not a common practice in engineering design, this approach is considered highly expensive for practical use.

The second approach for second-order plastic-zone analysis is based on the use of beam-column theory in which the member is discretized into line segments and the cross section of each segment is subdivided into FEs. Inelasticity is modeled considering normal stress only. When the computed stress at the center of any fiber reaches the uniaxial normal strength of the material, the fiber is considered to have yielded. Also, compatibility is treated by assuming that full continuity is retained throughout the volume of the structure in the same manner as in elastic range calculations. In the plastic-zone analysis, the calculation of forces and deformations in the structure after yielding requires an iterative trial-and-error process because of the nonlinearity of the load–deformation response and the change in cross-section-effective stiffness in inelastic regions associated with the increase in the applied loads and the change in structural geometry.

A plastic-zone analysis that includes the spread of plasticity, residual stresses, initial geometric imperfections, and any other significant second-order effects would eliminate the need for checking individual member capacities in the frame.

Therefore, this type of method is classified as advanced inelastic analysis in which the checking of beam-column interaction equations is not required. In fact, the member interaction equations in modern limit state specifications were developed, in part, by curve fit to results from this type of analysis. In reality, some significant behaviors such as joint's and connection's performances tend to defy precise numerical and analytical modeling. In such cases, a simpler method of analysis, which adequately represents the significant behavior, would be sufficient for engineering applications.

6.2.5 VERIFICATION OF PLASTIC-ZONE METHOD

Two full-size specimens of portal frame, Figure 6.5a, were tested for monotonic loading by Wakabayashi (Chen and Toma, 1994). The member sizes are summarized in Table 6.1, and the measured sectional and material properties are as given in Tables 6.2 and 6.3. The panels of the connections between columns and beams are stiffened with horizontal and diagonal plates to prevent shear buckling so that the bending of the members controls the ultimate strength of the frame. The specimens are supported in the out-of-plane direction at four equal points of the beam and at the midpoint of the columns by a hinge device that can rotate in the plane of the frame but is fixed out of the plane.

In the two specimens, all columns had section $H-175\times175\times7.5\times11$ mm and all beams had section $H-250\times125\times6\times9$ mm. The span length is 500 cm and the height is 260 cm. P is the column load, P_e is the elastic buckling load of loaded frame, P_y is the column yield load, h is the column height (260 cm), L is the beam length (500 cm), I_b and I_c are the moments of inertia of the beam and column, respectively, and r is the radius of gyration of the column.

In testing the specimens, no vertical load was applied to specimen FM0, but for specimen FM5, a constant vertical load was first applied on the top of the columns, after which the horizontal load at the top of the frame was increased gradually. The measured load-deflection curves are shown in Figure 6.5b and c for the two specimens, and some measured values are given in Table 6.4. The specimens contained imperfections, which caused early yielding, and the stiffness decreased at an early stage of horizontal loading. While specimen FM0, after reaching the mechanism, showed a gradual increase in horizontal load due to strain hardening, specimen FM5 showed stiffness degradation as the lateral deflection increased. The local buckling appeared not to affect much the horizontal load-carrying capacity.

The test results are compared in Figure 6.5b and c with the plastic-zone method and the plastic-hinge method. It is assumed in these analyses that both local buckling and lateral buckling do not occur and the displacements are small. In the plastic-zone method, the portal frame is assumed to be composed of two L-shaped frames by antisymmetry. Furthermore, the L-shaped frames are separated into two members, that is, a beam and a column. Applying the equilibrium and compatibility conditions at the joint of the beam and the column, the load–deformation relations are obtained. The members are divided into 25 segments. The rotation of the members is calculated by integrating the curvature along the length. The effects of shear force

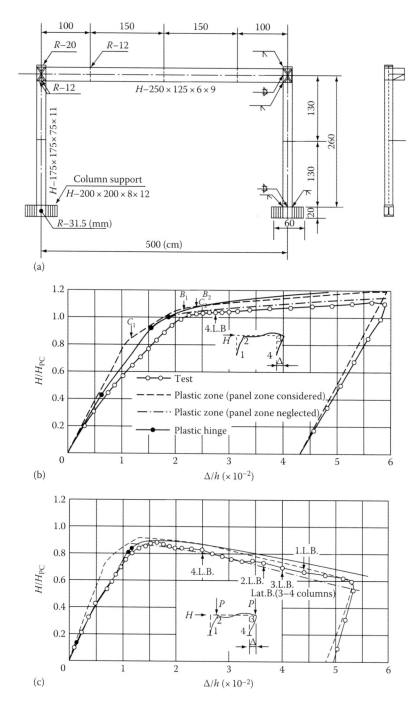

FIGURE 6.5 Full-size portal frames: (a) frame configuration; (b) horizontal force–displacement relation of specimen FM0; and (c) horizontal force–displacement relation of specimen FM5.

TABLE 6.1

Member Sizes and Test Program of the Full-Size Frames

Specimen Name	P (ton)	$\dfrac{P}{P_y}$	$\dfrac{P}{P_e}$	$\dfrac{h}{r}$	$\dfrac{I_b \times h}{I_c \times L}$
FM0	0	0	0	34.7	0.768
FM5	70	0.489	0.12	34.7	0.742

TABLE 6.2

Actual Section Properties of the Full-Size Portal Frames

Specimen Name	Column				Beam			
	A (cm²)	I (cm⁴)	Z (cm³)	Z_p (cm³)	A (cm²)	I (cm⁴)	Z (cm³)	Z_p (cm³)
FM0	48.8	2740	314	351	37.9	4050	325	367
FM5	50.6	2840	323	363	37.3	4050	322	363

Note: A is the cross-sectional area, I is the sectional inertia, Z is the section modulus, and Z_p is the plastic section modulus.

TABLE 6.3

Material Properties of the Full-Size Portal Frames

Specimen Name	Column					Beam				
	σ_y (ton/cm²)	σ_u (ton/cm²)	ε_u (%)	$\dfrac{\varepsilon_{st}}{\varepsilon_y}$	$\dfrac{E_{st}}{E}$	σ_y (ton/cm²)	σ_u (ton/cm²)	ε_u (%)	$\dfrac{\varepsilon_{st}}{\varepsilon_y}$	$\dfrac{E_{st}}{E}$
FM0	2.70	4.42	29.3	14.0	0.016	2.70	4.23	26.5	15.7	0.013
FM5	2.78	4.44	32.2	13.7	0.014	2.88	4.35	30.5	14.9	0.013

Note: σ_y is the yield stress, σ_u is the ultimate strength, ε_u is the maximum elongation, ε_y is the yield strain, ε_{st} is the strain at start of strain hardening, E is the modulus of elasticity, and E_{st} is the strain-hardening modulus.

and stiffness of the joint panel are considered. In Figure 6.5b, the yielded segments are indicated: for example, C_1 or B_2 means that yielding occurs in the first column segment or the second beam segment, respectively, from the member end. From the obtained sample results and from many other test results (Chen and Toma, 1994), it is obvious how the plastic-zone method can accurately predict the behavior of steel frames.

TABLE 6.4
Test Results of Full-Size Portal Frames

Specimen Name	H_i (ton)	Test Results			H_{pc} (ton)
		$\dfrac{H_f}{H_{pc}}$	$\dfrac{\Delta_f}{h}$	$\dfrac{\Delta_f}{\Delta_{fy}}$	
FM0	15.8	1.03	0.059	4.04	15.30
FM5	8.5	0.89	0.015	2.20	9.59

Note: H is the maximum horizontal force, H_f is the experimental, H_{pc} is the rigid plastic, Δ_f is the maximum displacement, and Δ_{fy} is the displacement at yield.

6.2.6 PRACTICALITY OF THE PLASTIC-ZONE METHOD

Whereas the plastic-zone solution is regarded as an exact solution, the method may not be used in daily engineering design because it requires very intensive computation efforts. Its application is limited to

1. Study detailed structural behavior
2. Verify the accuracy of simplified methods
3. Provide comparison with experimental results
4. Derive design methods or generate charts for practical use
5. Apply for special design problems

The AISC-LRFD beam-column equations were established in part based on a curve fit to the exact strength curves obtained from the plastic-zone analysis by Kanchanalai (Liew et al., 1993). The plastic-zone method will be used in this chapter as a benchmark for the development of the practical advanced analysis methods.

6.3 SIMPLIFICATIONS FOR ADVANCED ANALYSIS

6.3.1 INTRODUCTION

Due to the difficulty of applying the plastic-zone method in daily practice, the elastic–plastic hinge-by-hinge method, explained in Chapter 3, can be advanced in order to perform simplified advanced analysis. This can be achieved by accounting for the stability second-order effects, residual stresses, initial imperfection of members, inelasticity, and joint flexibility. In the following sections, simplified concepts are introduced in order to account for the aforementioned parameters in the elastic–plastic hinge-by-hinge method. These concepts are formulated in the next section in order to achieve a practical advanced analysis.

6.3.2 Stability Effects

The stability effects can be considered by writing the equilibrium equations of members in the deformed geometry, that is, by using the slope-deflection equations of beam-column to express member equilibrium. In these equations, the stability functions, as explained in Chapter 2, account for these second-order effects. This approach leads to significant saving in modeling and solution efforts by using one or two elements per member.

6.3.3 Residual Stresses

The tangent modulus concept of the Column Research Council (CRC) can be employed to capture the effect of residual stresses along the member between the plastic hinges. In this concept, the elastic modulus E, instead of the moment of inertia, is reduced to account for the reduction of the elastic portion of the cross section since the reduction of elastic modulus is easier to implement than that of moment of inertia for different sections. The reduction rate in stiffness between weak and strong axes is different, but this is not considered here. This is because a rapid degradation in stiffness in the weak-axis strength is compensated well by the stronger weak-axis plastic strength. As a result, this simplification will make the present method practical.

6.3.4 Initial Geometric Imperfection

There are two types of initial geometric imperfections of steel members: out-of-straightness and out-of-plumbness. These imperfections create additional moments in the column members, causing the member bending stiffness to be further reduced. Any of the following three methods can be used to model geometric imperfection: explicit imperfection modeling method, equivalent notional load method, and the further reduced tangent modulus method (Figure 6.6).

The ECCS (1984, 1991), AS (1990), and CSA (1989, 1994) specifications recommend the out-of-straightness varying in parabolic shape with a maximum in-plane deflection at the mid-height. It is noted that this explicit modeling method in braced frames requires the inconvenient imperfection modeling at the center of columns, although it is much lighter than that of the conventional LRFD method for frame design.

The ECCS (1984, 1991) and CSA (1989, 1994) also introduced the equivalent load concept (notional load concept), which accounts for the geometric imperfections in unbraced frames, but not in braced frames. The notional load method uses equivalent lateral loads to approximate the effect of member initial imperfection and distributed plasticity. This technique is similar to the concept of the "enlarged" geometric imperfection approach of the EC3 (2005), which is also allowed by the CSA (1994) and the AS (1990). In the ECCS, the exaggerated notional loads of 0.5% times the gravity loads are used to avoid overprediction of the strength of members as does the elastic–plastic-hinge method. The application of these notional loads is illustrated in Figure 6.7 to several frame examples. In the Eurocode 3, the equivalent notional load concept may also be used for braced frames, where this load may

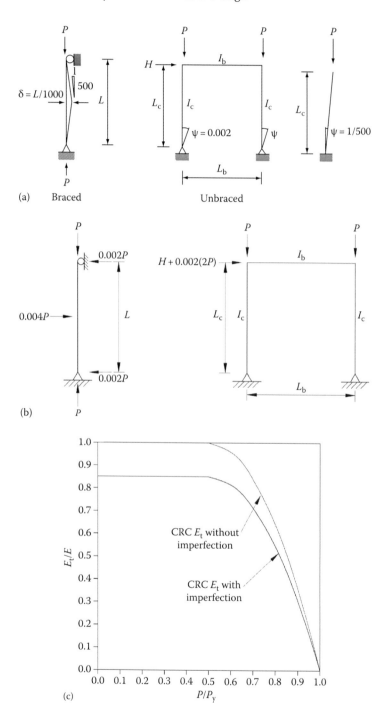

FIGURE 6.6 Geometric imperfection methods: (a) explicit imperfection modeling; (b) equivalent notional load; and (c) further reduced CRC tangent modulus $E'_t = 0.85E_t$.

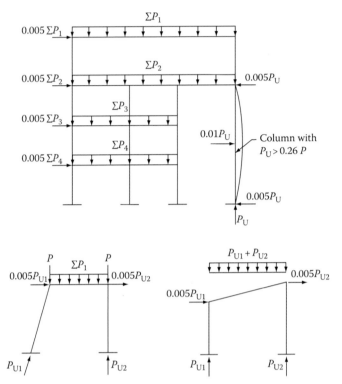

FIGURE 6.7 Examples of notional loads application for second-order elastic–plastic analysis.

be applied at mid-height of a column since the ends of the column are braced. For unbraced frames, the geometric imperfections of a frame may be replaced by the equivalent notional lateral loads expressed as a fraction of the gravity loads acting on the story. The drawback of this method for braced frames is that it requires a tedious input of notional loads at the center of each column, whereas the drawback of this method for unbraced frames is that the axial forces in the columns must be known in advance to determine the notional loads before analysis, which is often difficult to calculate for large structures subject to lateral wind loads.

In order to avoid the drawbacks of the previous two methods, it is recommended that the further reduced tangent modulus method be used. The idea of using the reduced tangent modulus concept is to further reduce the tangent modulus, E_t, to account for further stiffness degradation due to geometric imperfection (Figure 6.6c).

6.3.5 INELASTICITY

Material inelasticity can be simply introduced in the advanced analysis through the plastic-hinge concept. For cases with small axial forces and large bending moments, a gradual stiffness degradation of plastic hinge is required to represent the distributed plasticity effects associated with bending actions. Therefore, the *work-hardening plastic-hinge model* is adopted to represent the gradual transition from elastic stiffness to zero stiffness associated with a fully developed plastic hinge.

6.3.6 Joint Flexibility

Connections in real structures do not possess the idealized characteristics typically used in design codes around the world of being fully restrained (rigid) or simple (pin ended). The most commonly used connections in steel buildings are semi-rigid partial strength connections (partially restrained). Several studies have demonstrated the potential for improving stability, strength, and serviceability of structures with semi-rigid partial strength connections (Chen, 1993; Chen and Kim, 1998).

Connections may be modeled as a rotational spring with the appropriate moment–rotation relationship where an appropriate joint model can be adopted and the stiffness matrix of the member attached to the semi-rigid joint is modified accordingly. A comprehensive handbook on semi-rigid connections edited by Chen et al., including a list of collected connection test data in tabular form with illustrative figures for implementation, is available for engineering practice (Chen et al., 2011).

6.4 PRACTICAL ADVANCED ANALYSIS

6.4.1 Introduction

Since the plastic-zone method is not practical for everyday practice, the simplifications presented in the previous section are adopted in order to perform a practical advanced analysis for daily usage. In the following sections, we shall make further modifications and simplifications of the elastic–plastic-hinge method to improve its performance and at the same time to make it practical and work in engineering practice. These modifications are grouped into three categories: geometry, material, and connection. Details of these modifications can be found in the two doctoral theses (Liew, 1992; Kim, 1996) and their subsequent papers (see, e.g., Kim and Chen, 1996; Liew et al., 1993). This refined plastic-hinge method will be called the *practical advanced method for frame design* in what follows. In the following sections, the different parameters of this analysis are formulated based on these simplified concepts.

6.4.2 Stability Effects

In order to capture second-order stability effects, the conventional stability functions are used since they lead to a large saving in modeling and solution efforts by using one or two elements per member. With reference to Figure 6.8, the incremental force–displacement relationship of member ends may be written as:

FIGURE 6.8   Force–displacement of a beam-column element.

$$\begin{bmatrix} \dot{M}_A \\ \dot{M}_B \\ \dot{P} \end{bmatrix} = \frac{EI}{L} \begin{bmatrix} S_1 & S_2 & 0 \\ S_2 & S_1 & 0 \\ 0 & 0 & A/I \end{bmatrix} \begin{bmatrix} \dot{\theta}_A \\ \dot{\theta}_B \\ \dot{e} \end{bmatrix} \tag{6.4}$$

where

S_1 and S_2 are the stability functions

M_A and $\dot{M}_B$ are the incremental end moments

$\dot{P}$ is the incremental axial force

$\dot{\theta}_A$ and $\dot{\theta}_B$ are the incremental joint rotations

$\dot{e}$ is the incremental axial displacement

A, I, and L are the area, moment of inertia, and length of the beam-column element, respectively

E is the modulus of elasticity

The stability functions for in-plane bending of a prismatic beam-column, as illustrated in Chapter 2, are

$$S_1 = \frac{kL \sin kL - (kL)^2 \cos kL}{2 - 2\cos kL - kL \sin kL} \tag{6.5a}$$

$$S_2 = \frac{(kL)^2 - kL \sin kL}{2 - 2\cos kL - kL \sin kL} \tag{6.5b}$$

For simplicity in computation, Lui and Chen (1987) proposed approximate expressions for the stability functions S_1 and S_2 if $-2.0 \le kL \le 2.0$:

$$S_1 = 4 + \frac{2\pi^2 \rho}{15} - \frac{(0.01\rho + 0.543)\rho^2}{4+\rho} - \frac{(0.004\rho + 0.285)\rho^2}{8.183+\rho} \tag{6.6a}$$

$$S_2 = 2 - \frac{\pi^2 \rho}{30} + \frac{(0.01\rho + 0.543)\rho^2}{4+\rho} - \frac{(0.004\rho + 0.285)\rho^2}{0.183+\rho} \tag{6.6b}$$

where the nondimensional axial force P is defined as

$$\rho = \frac{P}{P_e} = \frac{P}{\pi^2 EI/L^2} \tag{6.6c}$$

and

$$k^2 = \frac{P}{EI} \tag{6.6d}$$

6.4.3 Residual Stresses

The CRC tangent modulus concept is employed to account for the gradual yielding effect due to residual stresses along the length of members under axial loads between two plastic hinges. Based on the CRC column strength formulas (Galambos, 1988), the tangent modulus can be written as

$$\frac{E_t}{E} = 1.0 \quad \text{for } P \le 0.5P_y \tag{6.7a}$$

$$\frac{E_t}{E} = 4\frac{P}{P_y}\left(1 - \frac{P}{P_y}\right) \quad \text{for } P > 0.5P_y \tag{6.7b}$$

where P_y is the squash load. Equations 6.7a and b are shown in Figure 6.6c.

6.4.4 Initial Geometric Imperfection

The further reduced tangent modulus method is based on further reducing the tangent modulus, E_t, to account for further stiffness degradation due to geometric imperfection. The further reduced tangent modulus approach was proposed by Chen and Kim (1997) to account for a geometric imperfection of 1/500 of the column length or story height. A reduction factor of 0.85 was determined from calibration with the almost exact plastic-zone solutions. The same reduction factor is used for both braced and unbraced structures, including both the out-of-straightness and the out-of-plumbness. It is used to further reduce the CRC–E_t value as given in the following equations, and as shown in Figure 6.6c in solid curve marked with imperfection, or simply $E_t' = 0.85E_t$:

$$E_t' = 0.85E \quad \text{for } P > 0.5P_y \tag{6.8a}$$

$$E_t' = 3.4\frac{P}{P_y}\left(1 - \frac{P}{P_y}\right)E \quad \text{for } P \le 0.5P_y \tag{6.8b}$$

6.4.5 Inelasticity

Inelasticity is accounted for through the work-hardening plastic hinge. The tangent modulus model in Equation 6.8 is suitable for $P/P_y > 0.5$, but is not sufficient to represent the stiffness degradation for cases with small axial forces and large bending moments. A gradual stiffness degradation of plastic hinge is required to represent the distributed plasticity effects associated with bending actions. Therefore, the work-hardening plastic-hinge model is introduced to represent the gradual transition from elastic stiffness to zero stiffness associated with a fully developed plastic hinge.

To represent a gradual transition from the elastic stiffness at the onset of yielding to the stiffness associated with a full plastic hinge at the end, a parabolic model simulating the gradual degradation of the element stiffness due to the plastification of the steel section is used as shown in Figure 6.9. The factor η, representing a gradual stiffness reduction associated with flexure, was proposed by Liew et al. (1993):

$$\eta = 1 \quad \text{for } \alpha \leq 0.5 \tag{6.9a}$$

$$\eta = 4\alpha(1-\alpha) \quad \text{for } \alpha > 0.5 \tag{6.9b}$$

In this model, α is the force state parameter obtained from the limit state surface corresponding to the member ends. The term α is based on the AISC-LRFD sectional strength curve and is expressed as

$$\alpha = \frac{P}{P_y} + \frac{8M}{9M_P} \quad \text{for } \frac{P}{P_y} \geq \frac{2M}{9M_P} \tag{6.10a}$$

$$\alpha = \frac{P}{2P_y} + \frac{M}{M_P} \quad \text{for } \frac{P}{P_y} < \frac{2M}{9M_P} \tag{6.10b}$$

where
   P and M are the second-order axial force and bending moment at the cross section
   M_P is the plastic moment capacity

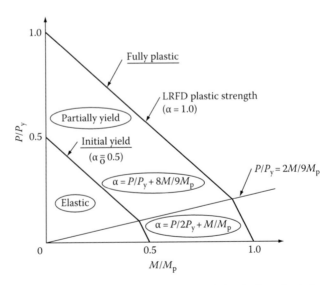

FIGURE 6.9 Smooth stiffness degradation for a work-hardening plastic hinge based on LRFD sectional strength curve as the limit or fully yielded surface.

Initial yielding is assumed to occur at $\alpha = 0.5$, and the yield surface function corresponding to $\alpha = 1.0$ represents the state of forces where the cross section has fully yielded and their corresponding state of forces (P, M) can move only along the yield surface under continuous loading condition.

6.4.6 INCREMENTAL FORCE–DISPLACEMENT RELATIONSHIP

In advanced analysis, the incremental load concept is applied to trace the force–displacement relationship of the structure and its components. When the work-hardening plastic hinges are introduced at both ends of a beam-column element, the incremental force–displacement relationship can be expressed as, in the usual notations,

$$\begin{Bmatrix} \dot{M}_A \\ \dot{M}_B \\ \dot{P} \end{Bmatrix} = \frac{E_t I}{L} \begin{bmatrix} S_{ii} & S_{ij} & 0 \\ S_{ij} & S_{jj} & 0 \\ 0 & 0 & A/I \end{bmatrix} \begin{Bmatrix} \dot{\theta}_A \\ \dot{\theta}_B \\ \dot{e} \end{Bmatrix} \tag{6.11}$$

where

$$S_{ii} = \eta_A \left[S_1 - \frac{S_2^2}{S_1}(1 - \eta_B) \right] \tag{6.12}$$

$$S_{jj} = \eta_B \left[S_1 - \frac{S_2^2}{S_1}(1 - \eta_A) \right] \tag{6.13}$$

$$S_{ij} = \eta_A \eta_B S_2 \tag{6.14}$$

where η_A and η_B are the stiffness reduction factors at ends A and B, respectively. Note that the modulus of elasticity E in Equation 6.11 is replaced by the tangent modulus E_t to account for the effects of residual stresses.

The parameter η represents a gradual stiffness reduction associated with flexures at sections. The partial plastification of cross sections at the end of elements is denoted by $0 < \eta < 1$. The parameter η may be assumed to vary according to the parabolic expression

$$\eta = 4\alpha(1 - \alpha) \quad \text{for } \alpha > 0.5 \tag{6.15}$$

The refined plastic-hinge analysis implicitly accounts for the effects of both residual stresses and spread of yielded zones. To this end, refined plastic-hinge analysis may be regarded as equivalent to the plastic-zone analysis.

6.4.7 Joint Flexibility

A connection may be modeled as a rotational spring with the appropriate moment–rotation relationship. Figure 6.10 shows a beam-column element with spring semi-rigid connections at both ends. If the effect of connection flexibility is incorporated into the member stiffness, the incremental element force–displacement relationship of Equation 6.11 should be modified as

$$
\begin{bmatrix} \dot{M}_A \\ \dot{M}_B \\ \dot{P} \end{bmatrix} = \frac{E_t I}{L} \begin{bmatrix} S_{ii}^* & S_{ij}^* & 0 \\ S_{ij}^* & S_{jj}^* & 0 \\ 0 & 0 & A/I \end{bmatrix} \begin{bmatrix} \dot{\theta}_A \\ \dot{\theta}_B \\ \dot{e} \end{bmatrix} \tag{6.16}
$$

where

$$
S_{ii}^* = \frac{\left(S_{ii} + \dfrac{E_t I S_{ii} S_{jj}}{L R_{ktB}} - \dfrac{E_t I S_{ij}^2}{L R_{ktB}} \right)}{R^*} \tag{6.17}
$$

$$
S_{jj}^* = \frac{\left(S_{jj} + \dfrac{E_t I S_{ii} S_{jj}}{L R_{ktA}} - \dfrac{E_t I S_{ij}^2}{L R_{ktA}} \right)}{R^*} \tag{6.18}
$$

$$
S_{ij}^* = \frac{S_{ij}}{R^*} \tag{6.19}
$$

$$
R^* = \left(1 + \frac{E_t I S_{ii}}{L R_{ktA}} \right)\left(1 + \frac{E_t I S_{jj}}{L R_{ktB}} \right) - \left(\frac{E_t I}{L} \right)^2 \left(\frac{S_{ij}^2}{R_{ktA} R_{ktB}} \right) \tag{6.20}
$$

where R_{ktA} and R_{ktB} are the tangent stiffness of connections A and B, respectively.

For modeling the connection, different models can be found in literature; however, one successful model is presented here, that is, the three-parameter power model by Kishi and Chen (1990). The model is expressed in terms of three parameters: (1) the initial connection stiffness R_{ki}; (2) the ultimate moment capacity of connection M_u; and (3) a shape parameter n. With reference to Figure 6.11,

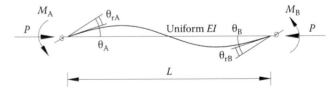

FIGURE 6.10 Beam-column element with semi-rigid connections.

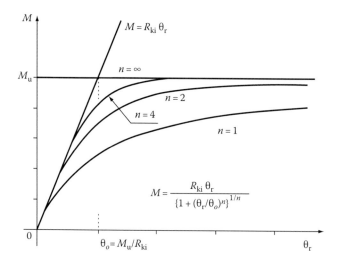

FIGURE 6.11 Moment–rotation behavior of semi-rigid connection three-parameter power model. (Modified from Kishi, N. and Chen, W.F., *J. Struct. Eng.*, *ASCE*, 116(7), 1813, 1990.)

$$m = \frac{\theta}{(1+\theta^n)^{1/n}} \quad \text{for } \theta > 0,\, m > 0 \tag{6.21}$$

where

$m = M/M_u$

$\theta = \theta_r/\theta_o$, where θ_o is the reference plastic rotation ($= M_u/R_{ki}$)

M_u is the ultimate moment capacity of the connection

R_{ki} is the initial connection stiffness

n is the shape parameter

When the connection is loaded, its tangent stiffness R_{kt} at an arbitrary relative rotation θ_r can be derived by simply differentiating Equation 6.21

$$R_{kt} = \frac{dM}{d|\theta_r|} = \frac{M_u}{\theta_o(1+\theta^n)^{1+1/n}} = \frac{R_{ki}}{(1+\theta^n)^{1+1/n}} \tag{6.22}$$

Also, from Equation 6.21,

$$\theta_r = \frac{M}{R_{ki}\left[1 - \left(\dfrac{M}{M_u}\right)^n\right]^{1/n}} \tag{6.23}$$

When the connection is unloaded, its tangent stiffness is equal to the initial stiffness:

$$R_{kt} = \frac{dM}{d\theta_r}\bigg|_{\theta_r=0} = \frac{M_u}{\theta_o} = R_{ki} \tag{6.24}$$

It is noted that R_{kt} and the relative rotation θ_r can be determined directly from Equations 6.22 and 6.23 without iteration, which makes the model an effective tool in design.

If the reader is interested in the semi-rigid frame design with LRFD using the advanced analysis method, the book by Chen et al. (1996) and Chen (2000) provide an in-depth coverage of the recent developments in the design of these frames. It also provides computer software that will enable the reader to perform planar frame analysis in a direct manner for a better understanding of behavior and a more realistic prediction of a system's strength and stability. Practical, analytical methods for evaluating connection flexibility and its influence on the stability of the entire framework are also described. These methods range from simplified member-by-member technique to more sophisticated computer-based advanced analysis and design approaches, including many example problems and detailed design procedures for each type of method.

6.5　APPLICATION EXAMPLES

6.5.1　Verification Example (Vogel's Six-Story Frame)

Vogel (1985) presented the load–displacement relationship of a six-story frame, Figure 6.12a, using plastic-zone analysis. The stress–strain relationship is elastic–plastic with linear strain hardening, Figure 6.12b, the residual stresses are as shown in Figure 6.12c, and the geometric imperfection in all columns is $L_c/450$, where L_c is the column length.

Vogel's frame has been analyzed using three different methods: the explicit modeling method, the notional load method, and the further reduced tangent modulus method. For comparison, an out-of-plumbness of $L_c/450$ is used in the explicit modeling method. A notional load factor of 1/450 expressed as a fraction of the gravity loads acting on the story, and the reduced tangent modulus factor of 0.85 are used. Since the further reduced tangent modulus is equivalent to a geometric imperfection of $L_c/500$, an additional geometric imperfection of $L_c/4500$ is modeled in the further reduced tangent modulus method, where $L_c/4500$ is the difference between the Vogel's geometric imperfection of $L_c/450$ and the proposed geometric imperfection of $L_c/500$.

The load–displacement curves of the three methods together with the Vogel's second-order plastic-zone analysis are compared in Figure 6.12d. The error in strength prediction by the three methods is less than 1%. Under service load, the three methods predict the lateral displacements within less than 3% of Vogel's exact solution, with the best prediction by the further reduced tangent modulus method.

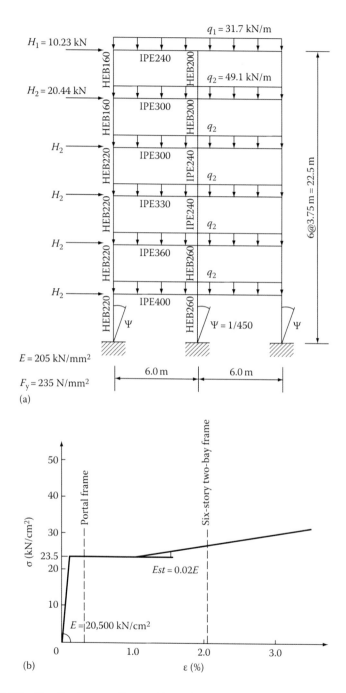

FIGURE 6.12 Vogel's six-story frame: (a) configuration and load condition; (b) stress–strain relationship;

(continued)

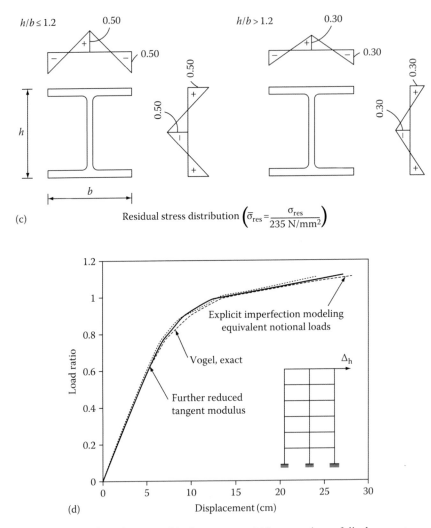

(c)   Residual stress distribution $\left(\bar{\sigma}_{res} = \dfrac{\sigma_{res}}{235 \text{ N/mm}^2}\right)$

(d)

FIGURE 6.12 (continued)   (c) residual stresses; and (d) comparison of displacement.

Thus, Vogel's frame is a good example of how the reduced tangent modulus method predicts well the frame response.

6.5.2   Design Example for Calibration against LRFD Code

Figure 6.13a shows a fixed-supported two-bay portal frame subjected to gravity and lateral loads with the critical load combination (Chen and Kim, 1997). All members are assumed to be laterally braced and are of A36 steel with preliminary sizes W16×89 used for beams and W14×68 for columns.

Each column is modeled as one element; the left beam has three elements whereas the right beam has two. In the equivalent notional load model, the notional load of 0.0306 kips results from 0.2% times the total gravity load of 153 kips and is added to the lateral load. In the further reduced tangent modulus model, the practical advanced analysis accounts for initial geometric imperfection by reducing the tangent modulus as explained in the previous section. The incremental load corresponding to each method is shown in Figure 6.13b and c; a scaling factor of 25.5 is used.

The load-carrying capacity from advanced analysis is shown in Table 6.5 for the two methods of initial member imperfection modeling. If the load-carrying capacity corresponding to the formation of the first plastic hinge is adopted, the preliminary

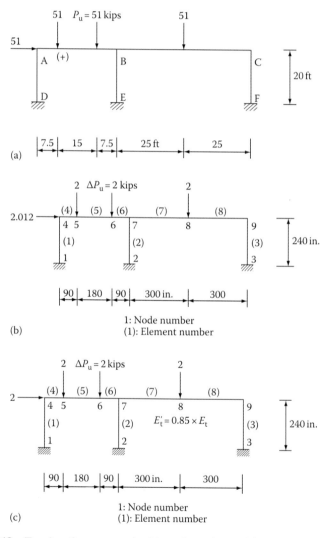

FIGURE 6.13 Two-bay frame example: (a) configuration and load condition of unbraced two-bay frame; (b) incremental load for the equivalent notional load model; (c) incremental load for the further reduced tangent modulus model;

(continued)

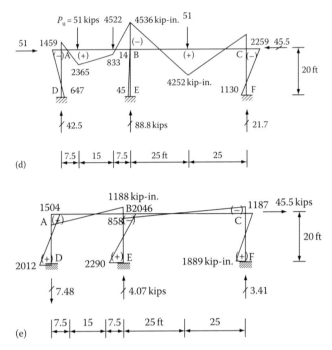

FIGURE 6.13 (continued)   (d) member forces from first-order analysis for non-sway case; and (e) member forces from first-order analysis for sway case.

TABLE 6.5
Load-Carrying Capacity of Two-Bay Frame Example

Imperfection Modeling	First Plastic Hinge	Ultimate State
Equivalent notional load	51.96	63.68
Further reduced tangent modulus	50.35	63.60

member sizes are suitable. On the other hand, if the ultimate state is selected, the member sizes can be reduced. However, other checks such as serviceability and ductility have to be carried out.

For design according to the AISC-LRFD method, the results of the first-order elastic analysis for the cases of non-sway and sway are shown in Figure 6.13d and e, respectively. Upon carrying out the design, it is found that the preliminary section sizes ($W16 \times 89$ for beams and $W14 \times 68$ for columns) are adequate.

When the load-carrying capacity corresponding to the formation of the first plastic hinge is considered to signify failure, the advanced analysis and the conventional LRFD method predict the same member sizes. On the other hand, if inelastic moment redistribution is considered (i.e., failure is characterized by the ultimate state), the method of advanced analysis may result in some saving in member sizes.

6.5.3 COMPARISON WITH THE LRFD DESIGN METHOD

The key considerations of the conventional LRFD method and the practical advanced analysis method are compared in Table 6.6. While the LRFD method accounts for key behavioral effects implicitly in its column strength and beam-column interaction equations, the advanced analysis method, as described previously, accounts for these effects explicitly in the form of stability function, stiffness degradation function, and stiffness reduction factor for geometric imperfections.

6.5.4 SEMI-RIGID FRAME DESIGN EXAMPLE

Connections in real structures do not possess the idealized characteristics typically used in design codes around the world of being fully restrained (rigid) or simple (pin ended). The most commonly used connections in steel buildings are semi-rigid partial strength connections (partially restrained). Several studies have demonstrated the potential for improving stability, strength, and serviceability of structures with semi-rigid partial strength connections.

Figure 6.14a shows a semi-rigid frame of two stories, each 12 ft high and one bay 25 ft wide. The frame is subjected to two loading conditions of distributed gravity and concentrated lateral loads, which are modeled as shown in Figure 6.14a (Chen and Kim, 1997). The roof beam connections are top- and seat-$L6 \times 4 \times 3/8 \times 7$ angle with double web angles of $L4 \times 3.5 \times 1/4 \times 5.5$ made of A36 steel. The floor beam connections are top- and seat-angles $L6 \times 4 \times 9/16 \times 7$ with double web angles of $L4 \times 3.5 \times 5/16 \times 8.5$. All fasteners are $A325 3/4''$ diameter bolts. The initial member

---

TABLE 6.6
Comparison of LRFD Method with Practical Advanced Analysis Method

Key Consideration	LRFD	Advanced Analysis
Second-order effects	Column curve, B_1, B_2 amplification factors	Stability function Necessary number of elements: • Beam with uniform load—2 • Column in braced frame—2 • Column in unbraced frame —1
Geometric imperfections	Column curve	Further reduced tangent modulus method • Reduction factor $E'_t = 0.85E_t$ • See Figure 6.6c
Stiffness degradation associated with residual stresses	Column curve	CRC tangent modulus curve • See Equations 6.7a and b
Stiffness degradation associated with flexure	Column curve, Beam-column interaction equations	Parabolic degradation function: • Refined plastic hinge concept (work-hardening plastic hinge) • See Equations 6.9 and 6.10 and Figure 6.9
Connection nonlinearity	No procedure	Power model/rotational spring

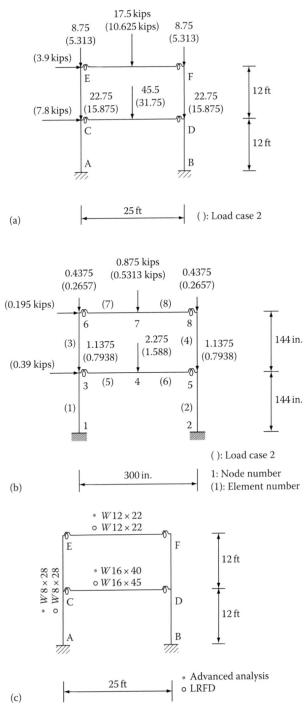

FIGURE 6.14 Semi-rigid frame example: (a) loadings, (b) load increment, and (c) sizes of sections.

TABLE 6.7

Estimated Parameters of Connections

Parameters	Connection at Roof Level	Connection at Floor Level
Initial stiffness R_{ki}	90 887 kip-in./rad	607 384 kip-in./rad
Ultimate moment M_u	446 kip-in.	1 361 kip-in.
Shape parameter n	1.403	0.927

sizes are selected as $W8 \times 28$, $W16 \times 40$, and $W12 \times 22$ for columns, floor beams, and roof beams, respectively.

Each column is modeled by one element and the beams are modeled by two elements. In the further reduced tangent modulus, the factor of 0.85 is used. The loads are divided into 20 increments. The joints are modeled by the three-parameter power model of Kishi and Chen. The parameters of the model are calculated, which are given in Table 6.7.

Upon performing the practical advanced analysis for load combinations 1 and 2, it was found that the ultimate load carrying at node 4, Figure 6.14b, is 44.8 and 39.2 kips, respectively. Compared to the applied loads of 45.5 and 31.75 kips, Figure 6.14a, the initial member sizes are adequate. However, when checking serviceability requirements, the drift due to lateral loads is found to be equal to 0.972 in., that is, span/296, which is greater than the limit, span/400. Therefore, the initial member sizes should be increased.

From the design according to the AISC-LRFD method, the column and roof beams remained the same as in advanced analysis. However, the floor beam is one size larger from the AISC-LRFD design since advanced analysis allows for inelastic moment redistribution, which results in smaller sections. The results from both methods are illustrated in Figure 6.14c.

6.6 PERFORMANCE-BASED DESIGN

Some drawbacks and limitations of the concept of effective length factor have been stated in Section 6.1.4. Aside from time saving and easiness in implementation in computer programming, a reliable representation of the interaction between the structural system and its members in a large structural system is guaranteed in advanced analysis. In addition, material saving, as a result of inelastic redistribution of internal forces in a structural system, can be achieved. Moreover, advanced analysis provides important information such as the failure mechanisms and progress of yielding of a structural system, which is critical in certain applications such as design for earthquakes.

In some applications such as performance-based design or design for earthquakes, the effective length factor is difficult to employ and it does not provide answer to some critical questions. On the other hand, advanced analysis is the solution in such a design. Besides, it can provide important information about which part of the structure is the most critical, and what will happen if part of the structure yields or fails. In addition, it illustrates how a structure will behave during an earthquake and what might happen if forces exceed the code limits. Moreover, it has become practical to design according to certain criterion such as the structure fuse concept, where failure is controlled at preselected locations, Figure 6.15, or performance-based design for different performances of structural components (Chen, 1999). In the preselected locations, section capacity is reduced, for example, by eliminating parts of the section flanges, such that yielding initiates at these sections. In this manner, plastic hinges are imposed at desired locations in order to avoid excessive cost and difficult repair of joints (Chen and Yamaguchi, 1996). With reference to Figure 6.15, it is realized that the preselected locations are easier to access and repair than joints in case of damage.

One of the applications of advanced analysis in performance-based design is the assessment of structure performance against fire. The failure mode of steel frames under fire and the length of time between initial introduction of fire and collapse can be determined from such an analysis (Hwa, 2003). From the standard fire curve,

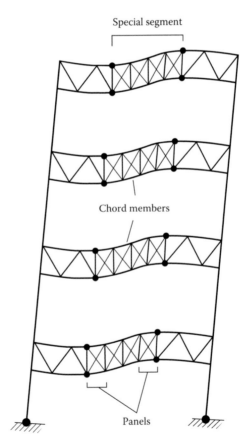

FIGURE 6.15 Special truss moment frame—an example of structural fuse concept.

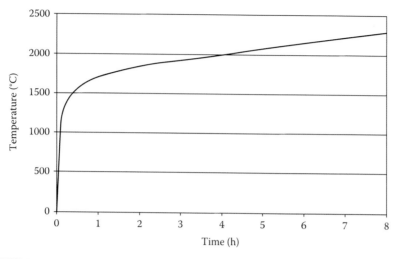

FIGURE 6.16 Time–temperature curve of standard fire.

for instance, the ASTM E119 (1998) curve shown in Figure 6.16, and based on the design fire duration, the design temperature is determined. Accordingly, the material properties, for example, elasticity modulus and yield stress, Figure 6.17, are modified for those members or parts of the structure affected by fire. Then, advanced analysis is performed under factored applied loads. Alternatively, the applied loads can be used in the analysis with the material properties of the zones affected by fire being modified due to increasing temperature values until collapse takes place. Then, the corresponding fire duration is determined.

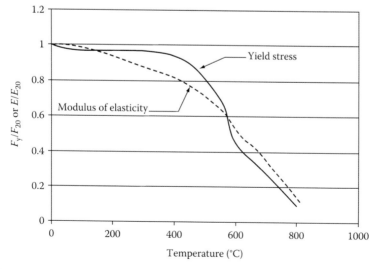

FIGURE 6.17 Effective yield stress and modulus of elasticity under elevated temperature.

6.7   HISTORICAL SKETCH

For decades, structural engineers and researchers have been exploring various approaches for assessing the structural behavior for the analysis and design of steel structures. The assessment methods, accordingly, have evolved over time, from hand calculation approach based on member capacity checks to computer-based approach based on advanced analysis to consider the interdependent effects between members and frame stability. The strength and stability of a structural system and its members are related, but this interaction has been treated separately in the current engineering practice by the popular effective length factor method. These indirect analysis and design methods include historically the ASD, PD, and LRFD. All these design approaches require an elastic analysis and separate specification member capacity checks including the calculation of the effective length factor, commonly known as the K factor.

With the present development of computer technology, two aspects, the stability of separate individual members and the stability of their structural frameworks as a whole, can be treated rigorously for the determination of the maximum strength of the structures. The development of this direct approach to design is called "advanced analysis" or more specifically, "second-order inelastic analysis for frame design." In this direct approach, there is no need to compute the effective length factor since separate member capacity checks encompassed by the specification equations are not required.

In the United States, the term "advanced analysis" strictly means "second-order inelastic analysis" for frame designs without the use of the effective length factor (K factor). There are three stages of progress to achieve the advanced analysis for frame design at present:

> *Stage 1*: Direct second-order elastic analysis eliminating the use of amplification factors. This direct computation of second-order forces was encouraged in the AISC-LRFD (1986) specifications. This is LRFD design but not advanced analysis for design.
>
> *Stage 2*: Direct second-order plastic analysis with the use of K factor to do member-by-member capacity checks with code requirements. This is a transition to advanced analysis, but it is still not advanced analysis because the K factor is still needed in the design process.
>
> *Stage 3*: Direct second-order inelastic analysis for frame design without the use of K factor to do member-by-member capacity checks with code requirements. The code requirements are met automatically in the advanced analysis for the structural system. This is called *advanced analysis*.

The purpose of this chapter is to present a simple, concise, and reasonably comprehensive introduction to a practical, direct method of steel frame design, using advanced analysis that will produce almost identical member sizes as those of the traditional LRFD method. The direct method described herein is limited to two-dimensional steel frames, so the spatial behavior is not considered. Lateral torsional buckling is assumed to be prevented by adequate lateral braces. Compact plastic W

sections are assumed so that sections can develop their full plastic moment capacity without buckling locally. All loads are statically applied.

Will the advanced methods result in buildings built with light members? Not necessarily. Will buildings cost less to design or build? Not likely. Will it be more complicated to design than LRFD? No! What it will do then? The advanced methods described here will encourage engineers to use more accurate analysis methods and computer programs. The advanced analysis can predict more accurately the possible failure modes of a structure, exhibit a more uniform level of safety, and provide a better long-term serviceability and maintainability. It is the state-of-the-art design methods for the structural engineers for the twenty-first century.

REFERENCES

AISC-LRFD, 1986, 1994, 1997, 2005, *The Load and Resistance Factor Design Specification for Structural Steel Buildings*, American Institute of Steel Construction, Chicago, IL.

AISC-LRFD, 2005, *Manual of Steel Construction*, American Institute of Steel Construction, Chicago, IL.

American Society of Testing and Materials (ASTM), 1998, Standard test methods for fire tests of building construction and materials, Report E119, PA.

Attala, N. M., Deierlein, G. G., and McGuire, W., 1994, Spread of plasticity: Quasi-plastic-hinge approach, *Journal of Structural Engineering*, 120, 8, 2451–2473.

Canadian Standard Association, 1989, *Limit States Design of Steel Structures*, CAN/CSA-S16.1-M89.

Canadian Standard Association, 1994, *Limit States Design of Steel Structures*, CAN/CSA-S16.1-M94.

Chen, W. F., ed., 1993, *Semi-Rigid Connections in Steel Frames*, Council on Tall Buildings and Urban Habitat, McGraw-Hill, New York, 318pp.

Chen, I. H., 1999, Practical advanced analysis for seismic design of steel building frames, PhD dissertation, School of Civil Engineering, Purdue University, West Lafayette, IN, 222pp.

Chen, W. F., ed., 2000, *Practical Analysis for Semi-Rigid Frame Design*, World Scientific Publishing Co., Singapore, 465pp.

Chen, W. F., 2009, Toward practical advanced analysis for steel frame design, *Journal of the International Association for Bridge and Structural Engineering (IABSE), SEI*, 19, 3, 234–239.

Chen, W. F. and Atsuta, T., 2007a, *Theory of Beam-Columns, Vol. 1 — In-Plane Behavior and Design*, McGraw-Hill, New York, 513pp., 1976, Reprinted by J. Ross, Orlando, FL, www.jrosspub.com

Chen, W. F. and Atsuta, T., 2007b, *Theory of Beam-Columns, Vol. 2 — Space Behavior and Design*, McGraw-Hill, New York, 732pp., 1977, Reprinted by J. Ross, Orlando, FL, www.jrosspub.com

Chen, W. F., Goto, Y., and Liew, J. Y. R., 1996, *Stability Design of Semi-Rigid Frame Design*, Wiley, New York.

Chen, W. F. and Han, D. J., 2007, *Plasticity for Structural Engineers*, Springer-Verlag, New York, 606pp., Reprinted by J. Ross, Orlando, FL, www.jrosspub.com

Chen, W. F. and Kim, S. E., 1997, *LRFD Steel Design Using Advanced Analysis*, CRC Press, Boca Raton, FL, 441pp.

Chen, W. F. and Kim, Y. S., 1998, *Practical Analysis for Partially Restrained Frame Design*, Structural Stability Research Council, Lehigh University, Bethlehem, PA, 82pp.

Chen, W. F., Kishi, N., and Komuro, M., eds., 2011, *Semi-Rigid Connections Handbook*, J. Ross, Fort Lauderdale, FL, Website: www.jrosspub.com

Chen, W. F. and Lui, E. M., 1986, *Structural Stability—Theory and Implementation*, Elsevier, New York, 490pp.

Chen, W. F. and Lui, E. M., 1991, *Stability Design of Steel Frames*, CRC Press, Boca Raton, FL.

Chen, W. F. and Lui, E. M., eds., 2005, Steel frame design using advanced analysis, in *The Handbook of Structural Engineering*, Chapter 5, 2nd edn., CRC Press, Boca Raton, FL.

Chen, W. F. and Sohal, I., 1995, *Plastic Design and Second-Order Analysis of Steel Frames*, Springer-Verlag, New York, 509pp.

Chen W. F. and Toma S., 1994, *Advanced Analysis of Steel Frames*, CRC Press, Boca Raton, FL.

Chen, W. F. and Yamaguchi, E., 1996, Spotlight on steel moment frames. *Civil Engineering, ASCE*, 66, 44–46.

ECCS, 1984, Ultimate limit state calculation of sway frames with rigid joints, Technical Committee 8—Structural Stability Technical Working Group 8.2—System, Publication No. 33.

ECCS, 1991, *Essentials of Eurocode 3 Design Manual for Steel Structures in Building*, ECCS-Advisory Committee 5, No. 65.

Eurocode-3, 2005, *Design of Steel Structures—General Rules and Rules for Buildings*, European Committee for Standardization, EN 1993-19-1, Brussels, Belgium.

Galambos, T. V., 1988, *Guide to Stability Design Criteria for Metal Structures*, 4th edn., Wiley, New York.

Hwa, K., 2003, Toward advanced analysis, in Toward advanced analysis in steel frame design, K. Hwa, ed., PhD dissertation, University of Hawaii, Honolulu, HI

Kim, S. E., 1996, Practical advanced analysis for steel frame design, PhD dissertation, School of Civil Engineering, Purdue University, West Lafayette, IN, 271pp.

Kim, S. E. and Chen, W. F., 1996, Practical advanced analysis for braced steel frame design, *Journal of Structural Engineering, ASCE*, 122, 11, 1266–1274.

Kishi, N. and Chen, W. F., 1990, Moment-rotation relations of semi-rigid connections with angles, *Journal of Structural Engineering, ASCE*, 116, 7, 1813–1834.

Liew, J. Y. R., 1992, Advanced analysis for frame design, PhD dissertation, School of Civil Engineering, Purdue University, West Lafayette, IN, 393pp.

Liew, J. Y. R., White, D. W., and Chen, W. F., 1993, Second-order refined plastic hinge analysis for frame design: Part I, *Journal of Structural Engineering, ASCE*, 119, 3196–3216.

Lui, E. and Chen, W. R. 1987, Steel frame analysis with flexible joints, *Journal of Constructional Steel Research*, Special Issue on Joint Flexibility in Steel Frames, 8, 161–202.

McGuire, W., 1992, Computer-aided analysis, in *Constructional Steel Design—An International Guide*, P. J. Dowling, J. E. Harding, and R. Bjorhovde, eds., Elsevier Applied Science, New York, pp. 915–932.

Salmon C. G. and Johnson J. E., 1990, *Steel Structure Design and Behavior*, 3rd edn., Harper Collins, New York.

Standards Australia, 1990, Steel structures, AS4100-1990, Sydney, Australia.

Vogel, U., 1985, Calibrating frames, *Stahlbau*, 10, 1–7.

White, D. W., 1993, Plastic hinge methods for advanced analysis of steel frames, *Journal of Constructional Steel Research*, 24, 2, 121–152.

White, D. W. and Chen, W. F., 1993, *Plastic Hinge Based Methods for Advanced Analysis and Design of Steel Frames: An Assessment of the State-of-the-Art*, Structural Stability Research Council, Bethlehem, PA, 299pp.

Ziemian, R. D. and McGuire, W., 1992, A method for incorporating live load reduction provisions in frame analysis, *Engineering Journal, AISC*, 29, 1–3.

7 The Era of Model-Based Simulation

7.1 THE ERA OF COMPUTER SIMULATION

In the past two decades, remarkable developments have occurred in the field of computer hardware and software. Advancements in computer technology have spurred the development of scientific simulation and visualization of problems, which traditionally have been addressed via experimentation and theoretical models. Such capabilities now offer the solution of problems thought "unsolvable" before, and are consequently, the reason behind the progress made in a number of areas in civil engineering. As a result of these rapid advances, a revolution has occurred in civil engineering research and practice. A third aspect, *computing*, has been added to the *theoretical* and *experimental* aspects of the field to form the basis of new civil engineering. Simulation plays an increasing critical role in all areas of science and engineering. Exciting examples of these simulations are occurring in areas such as *automotive crashworthiness* for component design in the auto industry, and *Boeing 777* for system design and manufacturing in aerospace.

The ongoing projects in space engineering include system design, assembly, and operation of the next generation Space Telescope (Hubble II), and the Accelerated Strategic Computing Initiative for system and component verification of selected focused applications for the U.S. Department of Defense, among others. In civil engineering, the U.S. National Science Foundation started an exploratory research program on model-based simulation (MBS) in 2000 and provided major funding for the establishment of the Network for Earthquake Engineering Simulation (NEES). The NEES is an interconnected system of earthquake engineering physical experimental facilities (e.g., shake tables, tsunami wave tanks, geotechnical centrifuges, reaction wall systems, and structural and geotechnical field facilities) in the United States. These component facilities distributed geographically around the nation will be networked to form a virtual integrated experimental facility, so that they can be used directly by the researchers in various locations to conduct large-scale experiments.

The exponential growth of computer speed and capacity has led to the development of computational science and engineering as a unique and powerful tool for scientific discovery. One key branch of this new discipline is *model-based simulation*, the objective of which is to develop the capability for realistically simulating the behavior of complex systems under loading and environmental conditions the systems may experience during their lifetime. Simulation does not replace observation and physical experimentation but complements and enhances their value in the synthesis of analytical models.

MBS is creating a new window into the natural and technological worlds. MBS is as valuable as theory and experimentation is for scientific discovery and technological innovation in that it provides a framework for combining theory and experimentation with advanced computation. Besides massive numerical computations, high-performance computers permit the use of other tools, such as visualization and global communications using advanced networks, all of which contribute to the ability to understand and to control the physical processes governing complex systems.

The engineering of civil and mechanical systems stands to benefit immensely from MBS because of their intrinsically complex nature, which involves materials that exhibit nonlinear behavior, multiple spatial and temporal scales, and different types of components whose behavior is governed by different physics. It is fascinating to envision that with the aid of MBS, it might be possible to understand from first principles phenomena, that today we can explain only crudely, the process by which a building or bridge collapses or how soil liquefies during an earthquake. This will naturally lead to improved design methodologies that are based on realistic simulations of performance, as compared with the current procedures in which performance is implicit and poorly understood. With designs based upon models, engineers can run high-fidelity simulations to evaluate new materials, components, and systems during the design stages before investing valuable resources in construction.

Moreover, civil and mechanical systems, in general, and structural and geotechnical systems, in particular, are associated with large uncertainties concerning their material properties and applied loads. By monitoring the system behavior, MBS can be used in conjunction with inverse methods to reduce the level of uncertainty by better identifying the material characteristics, the loading environment, and even adapting the model itself, thereby allowing designers to considerably increase the reliability of the system to achieve specified performance goals. In addition, MBS and the monitoring of the behavior of a complex system during construction, such as a large airport facility or a building complex, can guide the engineer in making real-time decisions concerning changes in the design or construction process. One can project that in the future, models will live concurrently with the systems they represent to provide owners and operators ongoing information that can be used in increasing operations and future performance.

7.2 MODEL-BASED SIMULATION IN STRUCTURAL ENGINEERING

Computer simulation has now joined theory and experimentation as a third path for engineering design and performance evaluation. *Simulation is computing, theory is modeling, and experimentation is validation* of the result. MBS is based on the integration of mechanics, computing, physics, and materials science for predicting the behavior of complex engineering and natural systems. MBS allows engineers and researchers to investigate the entire life cycle of engineered systems and to assist in decisions on the design, construction, and performance in civil and mechanical systems. Reliable and accurate MBS tools will permit the design of engineering systems that cost less and perform better. MBS promises to reduce design-cycle times while increasing system life span. MBS is currently being used effectively in some sectors of the engineering profession, notably in aerospace and automotive applications. The challenge is to make MBS a reality in other engineering fields.

The emerging areas of MBS in structural engineering will include, most notably, the following topics:

1. From the present structural system approach to the life-cycle structural analysis and design covering construction sequence analysis during construction, performance analysis during service, and degradation and deterioration analysis during maintenance, rehabilitation, and demolition
2. From the present finite element (FE) modeling for continuous media to the finite block types of modeling for tension-weak materials that will develop cracks, and subsequently, change the geometry and topology of the structure
3. From the present time-independent elastic and inelastic material modeling to the time-dependent modeling reflecting material degradation and deterioration science

A busy, productive, and exciting future lies ahead for all who wish to participate in these emerging areas of research and application. These areas are inherently interdisciplinary in science and engineering, where computation plays the key role. Scientists provide a consistent theory for application, and structural engineers must continue to face the reality dealing with these theories in order to make them work and applicable to the real world of engineering.

7.3 MBS SYSTEM INTEGRATION

The development of MBS for any civil engineering facility must involve the following steps: mathematical modeling, solution algorithm, and software development. These steps are illustrated in the following sections.

7.3.1 MATHEMATICAL MODELING

These models must be developed on the basis of mechanics or physics, and materials science based on the observation of experimental testing. The level of refinement for a mathematical model depends on its application. For example, to model a reinforced concrete column subject to an earthquake loading in a highway bridge, a plastic-damage continuum mechanics representation of damage at a macroscale may be adequate for a representation of the state of concrete material prior to the peak load under multiaxial stress conditions. However, in the post-peak load range near failure, the concrete material enters the strain-softening state, and localization occurs. The application of fracture mechanics to concrete, including the bond slip between rebars, and concrete at a microscale is necessary for an accurate representation of the failure process in the reinforced concrete column during its entire service life.

7.3.2 SOLUTION ALGORITHM

As mentioned previously, for an integrated life-cycle simulation of constructed facilities, it is not uncommon for realistic simulations to require the mathematical modeling of radically distinct scales, in time and/or space. For example, in the analysis of a reinforced concrete bridge system under seismic loading, a macroscale is necessary to model the overall behavior of the structure–soil system, whereas a microscale is needed for the fracture mechanics analysis at the local level in order to trace the development and propagation of microcracks and bond slip between rebars and concrete material. Therefore, for a realistic simulation of the behavior of these structures, the microcracking and bond slip must be treated at a microlevel. This behavior must then be "translated" to the level of the structure or the macrolevel. To this end, parallel numerical algorithms must be used, and a suitable interface algorithm for different scales must be developed.

In using parallel FE analyses for the structural system or parallel fracture mechanics analyses for the local microdamages of materials, a single initial partition of the physical domain is usually not sufficient to provide a good load balance in the parallel computations. This is because, in these analyses, some parts of the mesh behave nonlinearly, while others remain predominantly in the linear range. The computational effort involved in each of these parts of the mesh may be significantly different. Major improvements in computational efficiency can be gained in these cases by repartitioning the domain during the analysis to ensure that the workload in each processor remains roughly the same. Therefore, to successfully realize these simulations, proper algorithm design and analysis in a parallel or distributed environment must be carried out.

7.3.3 SOFTWARE DEVELOPMENT

To bring advanced computing and simulation capabilities to civil engineering applications, it is necessary to develop a domain-specific software development environment to support these types of focused engineering applications. A software development environment is a compatible set of tools, usually based on a specific software development methodology that can be employed for several phases of software and operation. There are hundreds of such systems on the market, and more are becoming available each year.

The key to domain-specific software development environments is software reuse. Software reuse enables the knowledge obtained from the solution of a particular problem to be accumulated and shared in the solution of other problems. If software accumulated from previous software development can be utilized in the development of new applications, substantial applications can be built more efficiently. This is an ideal environment for university research and education. Software reuse promises substantial improvement in several aspects: software productivity, maintainability, portability, standardization, and general quality. A flexible software development environment for a specific domain is essential to the sharing of software as well as to the proper adaptation of existing software to the new hardware environment.

Changes in computer hardware have spurred the investigation of new algorithms for engineering computation. For example, much research is currently being conducted on algorithms, which take advantage of parallel computing architectures. An ideal software development environment should facilitate the incorporation and testing of new algorithms in different application areas. The envisioned environment has a potential to greatly increase the rate of technology assessment and assimilation by automating parallel programming practices as well as software engineering that facilitates reuse.

Also, the envisioned software development environment is essential to permit the rapid prototyping of algorithms and the approaches needed for effective research. Parallel and distributed computing, computing graphics, advanced user-interface, and intelligent database should not only be investigated in fundamental research but should also be used as tools by researchers.

7.4   MATERIAL MODELING

In order to model failure and limit states, models of materials must represent a broad range of conditions. There are three "multi" axes on which research issues connected with material modeling can be identified: multiconstituent, multiscale, and multiphysics. These are characterized by the following:

- *Multiconstituent* material modeling recognizes that many substances of engineering importance possess complex heterogeneous component structures. This includes the aggregation of multiple components into an inhomogeneous form as well as the more well-understood cases of anisotropy and the nonlinear behavior of materials modeled as homogeneous.
- *Multiscale* material modeling recognizes that the behavior of a real material is governed by physical phenomena that arise from a wide range of geometric and temporal scales, ranging from atomic-force interactions in molecular dynamics to microstructural effects in localization, to the large-scale continuum response generally considered in the current MBS efforts in engineering. Similarly, different timescales are needed to represent material failure, which may initiate in microseconds, the overall system behavior under severe dynamic loads on the order of minutes, to a timescale of decades when simulating aging and deterioration of materials or earthquake processes. Finally, different scales may be necessary when integrating model simulation with advanced visualization aided by haptic devices. (Force feedback with haptic devices must be scaled to provide a meaningful interpretation of forces.) The consideration of time, length, and force scales in simulation models also provides a useful framework for conducting scaled model experiments on components and systems. Since full-scale testing is impractical, scaled experiments must satisfy the laws of similitude. When developing advanced models, it is advantageous to consider the scale issues not only for the simulation but also for designing experiments to validate the models.

- *Multiphysics* material modeling recognizes that it is no longer sufficient to merely characterize the single-physics response of an engineering material. Instead, the full web of interacting responses must be identified and reliably characterized if a fully coupled multiphysics simulation of a complex engineering system is to be resolved.

An additional issue alluded to the earlier issues is that of modeling interfaces between components. The most important physical behavior that affects the performance of engineered systems occurs at the interfaces between different components. The components may have very different physics, and the model of the interfaces must account for proper conservation laws and relationships between the components. There are numerous examples, but one particularly complex one is the representation of the flow of a liquefied soil around a large embedded foundation of a structure. In addition to modeling liquefaction, the flow-induced forces at the interface of the foundation must be coupled with a structural model.

7.5 INTEGRATION OF HETEROGENEOUS MODELS

High-fidelity simulations of civil and mechanical systems require the integration of different types of models, as discussed before. In civil engineering, an example is that of modeling a structure and its foundation and their surrounding environment that imposes loads and deformations on the structure. Depending on the context, the environmental loading may be wind, modeled as a fluid domain, or an earthquake propagating through the soil, modeled as a solid or solid–fluid domain. Another example is that of simulating the deformation capacity of a ductile connection in a steel-framed building subjected to low-cycle fatigue. The representation of the failure mechanism begins by modeling the fatigue processes at the microscopic scale. The results of these simulations are then used to determine the effects on the structural performance and safety at the macroscopic scale.

There are many other examples that illustrate the need for heterogeneous models. One such example is the wind, seismic, and life-cycle (e.g., traffic) simulation of a large bridge, which involves a plethora of materials, structural and geotechnical aspects, and seismological aspects, such as spatially variable seismic excitation, uncertainty, etc. Another example is the construction of a large building with a deep excavation in which MBS can play an important role in design and in speeding up construction. A different kind of application is the simulated cyclic testing of a steel-moment frame joint, the failure of which is not yet possible to predict reliably. Such a simulation world requires a multiscale modeling and probabilistic approach and could well serve as a benchmark for an experiment design within NEES. The output of the said simulation would be a distribution of failure modes. These examples, which are neither prescriptive nor exhaustive, are cited as a way of illustrating the range of applications that could benefit from MBS and to indicate that most of them require the participation of interdisciplinary teams.

In general, advanced simulation models share certain characteristics and differ in others, as detailed in the following:

- The three "multis," perhaps at different levels of abstraction and complicated interfaces, are inherent characteristics of advanced simulation models. After synthesizing and validating individual models, the challenge in computing with such heterogeneous models is to integrate them into a global model for the simulation of a complete system. The current computer software for simulation is typically "closed" in the sense that individual models that are available are contained within the software ("the code") and are implemented to be internally consistent. Simulation models of interest to CMS systems are partially based on the finite element method (FEM) and the boundary element method (BEM) spatial discretization. However, there may be other suitable modeling approaches, particularly for different length scales.

- The deployment of commercial software, coupled to improve pre- and post-processors, has considerably simplified the specification of complicated geometries. However, work remains to be done in robust 3D mesh generation and especially with regard to linkage to CAD/CAM. There is substantial ongoing research and development in geometric modeling and mesh generation. Integration of geometric models with other activities, such as prototyping, design, and manufacturing, is ongoing in both industrial and laboratory settings.

- The simulation is less clear as regards models, loads, and environment. For marketing reasons, commercial FEM and BEM software address lowest denominator "generic" needs. Their black box nature can make it difficult to find out what is actually implemented, hampering credibility and authentication. Most FEM and BEM programs allow inclusion of additional elements or materials. However, this is too low level for effective integration of heterogeneous models. With traditional software architecture, it is nearly impossible to integrate different models of subsystems and components, particularly if specialized solution strategies are needed for the models.

In order to advance MBS, there is an urgent research need for an open, integrative approach to modeling. The research issues are to define the inherent characteristics of simulation models and agree upon definitions of software interfaces for models. The software interfaces would include the physical interfaces of the models, with appropriate physics requirements that must be satisfied at the interface. The software is to be enriched by information about scale, complexity and level of abstraction, computational demands, and communication needs (including response data and visualization) for utilizing the models. In information technology terms, the characterization of a simulation model is meta-model (information about the model). With a flexible standard for meta-models, it is possible to build simulation frameworks with interchangeable parts. An engineer conducting a simulation within such a framework can draw upon a variety of very different models because the framework can obtain information about the models and compute with models through the meta-model description and software interfaces.

With the concept of meta-models, the use of configurable, distributed simulations using Internet technology becomes a possibility. Simulation frameworks would allow engineers to select appropriate models from the Internet (perhaps with Web site providing parameterized models) based on credibility requirements, the design stage of the system, and the computational resources available. Within the distributed framework, the component and material models are integrated into a system model. Computing resources available on the network (in the form of a computational grid) are used to perform the simulation. Finally, visualization and data mining (possibly distributed) technology would be used to interpret and interrogate the simulation.

The vision of MBS is radically different from the current use of simulation "codes." The development of all-encompassing codes that are accessible and usable by engineers has probably reached a limit. The future of MBS lies in research to move to a true distribution of not only equation solving but also model-building, simulation, and visualization interpretation. This will provide much more flexibility and room for innovation in using models for simulation that is possible with current codes.

Achieving flexible integration requires the development of standards for the meta-models and the interoperability of models. Emerging standards on meta-data (such as XML) and data exchange for scientific application and application interoperability (such as COM and CORBA) must be pursued for MBS. Ultimately, we envision that industry-wide standards will lead to open frameworks for exchanging and synthesizing simulation models and allow the selection of computational resources on a "computational grid." Civil and mechanical researchers should be cognizant of standardization efforts in other fields so as to accelerate the integration of simulation models with emerging technology.

7.6 REPRESENTATION AND PROPAGATION OF UNCERTAINTY

The models of civil systems must represent the large uncertainty entitled in characterizing the models, material parameters, and the environmental loads. Uncertainty arises from a number of sources: lack of data on material properties, particularly for soils; inherent material variability, both initially and due to deterioration over time; construction processes; and the randomness of the loading environment. There is additional uncertainty associated with the selection of a particular model for a physical process. Since perfect models do not exist, there is an error associated with the model itself. Finally, in dealing with civil systems, the uncertainty in human decision making must be recognized. There are numerous cases in which the failure of a system could be traced to decisions concerning design, construction, or operation.

The challenge in MBS is the representation of uncertainty in a complete and consistent manner within a simulation. The modeling of uncertainty and propagation of uncertainty from model parameters to a probabilistic estimate of performance involves research challenges when dealing with complex systems of different materials and components. Research needs include efficient methods for computing the sensitivity of system performance to changes in model parameters, probabilistic characterization of model properties, methods for updating models with new data to reduce uncertainty, and integrating component reliability to provide estimates of system reliability. It should be noted that sensitivity of simulated responses is an important ingredient for optimization and model updating, in addition to evaluating system reliability. Finally, there is a research need to represent human decision-making processes, and attendant uncertainties in those processes, on system performance. A complete simulation of performance needs to include human factors.

7.7   MODEL SYNTHESIS

The choice of analytical and numerical models of an engineering system is an essential part of MBS. Such a decision represents a delicate balancing act. The model must be able to capture the essential physics while deliberately ignoring aspects irrelevant to engineering decisions. Models may be distinguished by their level of abstraction of the physics in the system. An engineer may wish to consider several models simultaneously, each at a different level of abstraction. High-level abstractions may be appropriate for exploring alternative designs and strategies early in the process. The models become progressively refined as design decisions harden. Moreover, the models may exist over the life of the system, for monitoring and retrofit purposes, in which case the models are updated using actual performance and response data. Ultimately, simulation models must be properly verified to serve a credible design and decision tool.

The synthesis of a simulation model requires selection, parameter identification, updating, and validation. Selection requires decisions about modeling approach, level of abstraction, and computational requirements. Parameter selection must deal with the use of experimental data and its uncertainty. Model updating involves reducing the uncertainty of the simulations either by reduced uncertainty in the data or by improving the model itself. Validation is the process of confirming that the model provides a reasonably complete and accurate picture of the actual physical behavior of the system.

There is a vast array of analytical models that have been developed over the past two centuries to represent material behavior and loads on civil systems. Their accessibility status for MBS use varies from immediate to difficult. At the easy-access end lie the classical macroscopic models of continuum mechanics of common structural, nonstructural, and geotechnical materials. Some programs include digitized tables in terms of important constitutive variables, such as temperature, stress level, and age, which are simply accessible on giving a name. It is relatively easy to incorporate this type of data into a simulation and to account for uncertainty.

Model synthesis is more difficult for more complex materials, for example, anisotropic media, composites, aggregates, foams, granular, and multiphase media. There are also phenomenological models that represent discrete components, such as joints and interfaces as well as localization, aging, fatigue, and progressive damage. Going down the length scale, there is increasing knowledge of material behavior at the meso- and microscale derived from discrete particle, crystal, and molecular models. In geotechnical applications, the selection of soil models is strongly linked to site characterization as well as (for seismic analysis) geophysical data. A similar knowledge fragmentation is evident for the load model.

The criteria for an acceptable model may depend on the simulation application. In some cases, system constraints may dictate that the system remains within its linear range of behavior under operating conditions. For instance, a facility designed for the fabrication of highly precise components with small tolerances, such as computer chips, requires that the maximum vibrations of the structure and its foundation be severely restricted. On the other hand, other CMS systems, such as building structures or dams designed to resist earthquakes, allow a certain amount of damage. For breakthroughs to occur in the MBS of systems that can be expected to behave non-linearly under their design loads, the models should be able to simulate limit states and failure. Other characteristics that models should possess are

- Ability to support simulation of fabrication and construction processes with "what if" scenarios. Use of MBS for such purposes is routine in aerospace projects, such as ongoing Space Station assembly, but less common in CMS.
- Incorporation of uncertainty.

7.8   COMPUTING

With the aid of high-performance computing, future structural analysis should be able to simulate the whole life cycle of structures through design, construction, service, degradation, and failure. An integrated life-cycle simulation of constructed facilities should be a focused and interesting topic for the development of the needed technologies for the MBS effort in civil engineering applications. The visualization of the virtual construction process of an oil and gas offshore platform, the modeling of an entire life cycle of a tall building, or the simulation of a bridge collapse during a strong earthquake are just a few examples of the potential impact that MBS development might have on civil engineering practice.

Advanced computer technology provides the possibility of speeding up the pace and widening the range of civil engineering research on simulations. However, the tremendous increase in computing power offered by modern high-performance computing cannot be fully and readily utilized by engineers due to the difficulties encountered in maintaining existing software and developing new software. Efforts on software development cannot be accumulated, which represents a great waste of specific and research resources, especially for upcoming MBS developments for large-scale civil engineering applications. Software development and maintenance are the main barriers, among others, in the infusion of advanced computer technology in civil engineering applications. Research is, therefore, urgently needed to apply the principles and methodologies of software engineering to civil engineering computing to overcome these barriers and to meet the increasing simulation demands of civil engineering research and instruction on software. An integrated domain-specific software development environment may provide the solution.

There is considerable room for improvement in the integration of computer science techniques and information technology into the computational models used in mechanical and civil engineering practice. These proven technologies include the following:

- Better use of parallelization for achieving scalable performance gains with parallel computing architectures. Most current commercial applications used in the profession are still serial, and, hence, cannot solve many important problems with the desired level of resolution. Using proven existing and feasible near-term software engineering techniques for shared- and distributed-memory computer architectures will substantially broaden the scope of civil and mechanical engineering problems that are amenable to computational simulation.

- Adaptive refinement of computed solutions, using a posteriori error measure, promises the dual reward of more rational use of computational resources (by only improving the computed solution where underlying physics warrants increased effort) as well as a more information-rich model-generated output stream for visualization and postprocessing. Further research on adaptivity, load balancing, and error-estimate-driven solution techniques pertinent for use in these professions is a key component for realizing these twin goals.

- Improved analytic formulations will be needed for handling the full complexity of multiphysics simulations in engineering. Research will be needed on techniques, such as arbitrary Lagrangian–Eulerian (ALE) discretization, unconditionally stable temporal operators for nonlinear formulations, and other new computational and analytic technologies for handling the full range of complex physical response found in engineering practice.

- Support of distributed computing architectures (e.g., distance computing, geographically disparate research teams, and network and communications infrastructure issues) is essential to support future needs of the civil and mechanical engineering professions.

- Data management and abstraction tools, such as data mining, will be required to wade through the immense volumes of data generated by large-scale simulation techniques, especially when those techniques are used in a design–analysis cycle with iterations required to handle the complex constraints of typical engineering practice.

7.9 SCIENTIFIC VISUALIZATION

Humans have five senses, but only the sense of sight has sufficient bandwidth to permit conveyance and interpretation of the immense amount of data obtained from large-scale simulation applications. This simple physiological constraint naturally leads to the use of scientific visualization tools to help sort through the bewildering amounts of data resulting from typical MBSs.

The full gamut of visualization tools are required for the complete integration of computational simulation into engineering practice, including animation for time-dependent (or other parametric) results; field-based display schemes for the usual scalar-/vector-/tensor-valued solution fields generally found in engineering analyses, and the use of color and other cues to add further dimensions to the resulting displays. Windowed and immersive virtual-reality techniques can aid engineers during the various components of the design-analysis cycle, and are naturally seen as the visual adjunct of the larger virtual physics setting that underlies the use of MBS in the mechanical and civil engineering disciplines.

Visualization methods must allow engineers not only to investigate the simulation of one model but also to compare multiple simulations for different models. An engineer or researcher would typically use several models to investigate the performance of a system, with different choices of parameters (or distributions), modeling techniques, or level of modeling abstraction. Visualization is the key mechanism to exploring behavior represented by a model and to identifying differences between models.

7.10 MODEL UPDATING AND VALIDATION

Analytical models generally contain parameters in the material and load models as well as variances in geometric dimensions. Since the parameters may have large uncertainties, as discussed before, the usefulness of a simulation may depend on the ability to update the model with measured or experimental data. Model updating encompasses methods that use sensitivity and optimization techniques to adjust those parameters and sizes so that the discrete model matches the observed properties, such as mode shapes and vibration frequencies. Those properties are obtained by measurements of the nondestructive type.

Model updating is a restricted form of the inverse problem, also called system identification. When applied over the lifetime of a system, it can become a technique for damage or failure detection (sometimes referred to system as "health monitoring"). This aspect is gaining increasing attention because of concern over the maintenance and repair of the aging U.S. infrastructure. Note that model updating only changes free parameters but not the model selection.

Model updating should not be confused with model validation, which involves a more comprehensive and critical study of the predictive capability of different models. Validation addresses the credibility of model predictions. It is done by comparing the simulated physical behavior against a comprehensive set of experiments on, or observation of, actual systems. The notion of MBS-predictive credibility is one of the key barriers that needs to be to overcome in the use of MBS techniques by practicing engineers, particularly in civil systems and geotechnical applications. Consequently, model validation is likely to emerge as a key research issue.

A logical approach to dealing with model validation is a tight linkage between simulation research and experimental research. Simulation models may be used to design experiments needed to validate the simulations. In a well-coordinated program, key simulations are used to identify experiments needed to test hypothesis revealed in the simulations, and experiments are then used to validate the models and hypotheses. Whenever possible, this process ought to be ultimately supplemented by comparing predictions from the simulations with observations of the response of actual systems.

For model verification, it will be essential that researchers have access to comprehensive (raw) measurements, interpolated as necessary to collocate with discrete model locations (such as node points in an FEM or BEM model) as well as documentation and authentication of the experimental method. This is particularly important for model updating and life-cycle system identification. For geotechnical applications, new satellite-based remote-sensing technology can provide outstanding ability to collect large amounts of geophysical data for validating models of soil and geotechnical systems.

7.11 SUMMARY

MBS is inherently interdisciplinary in science and engineering, in which computation plays a fundamental role. The process of integration involves several basic steps starting with physical measurements or experiments as the basis for the development of relevant mathematical models for a physical system followed by the design of a proper algorithm for its numerical solution implementing the procedure with the necessary coding and software interface development, selecting a proper hardware to run the computer simulation of the physical system, validating the computer model with physical testing, providing graphical visualization of the simulated results, and sharing the simulation model with others through high-speed network communication.

The benefits of MBS development for engineering practice are as follows:

1. Simulation enables a more thorough exploration of design space at a much lower cost.
2. Simulation is an efficient tool to reduce product cost.
3. Simulation provides a faster turnaround of a product cycle.
4. Simulation is interdisciplinary and is good for a broad-based engineering education for lifelong learning.

MBS certainly cannot replace tests completely, but at least can minimize the use of physical tests. In fact, it can best be used to maximize the value of physical tests with computer simulation.

SELECTED RELEVANT REFERENCES

Chen, W. F., 2000, Model-based simulation in civil engineering: Challenges and opportunities, plenary address, *Symposium on the Applications of Electronic Computing in Civil and Hydraulic Engineering*, Taichung, Taiwan, February 17–18, pp. 1–15.

Chen, W. F., 2009, Seeing the big picture in structural engineering, *Proceedings of the Institution of Civil Engineers*, 162, No. CE2, 87–95.

Chen, W. F., Sotelino, E. D., and White, D. W., 1993, *High Performance Computing in Civil and Hydraulic Engineering*, National Cheng-Kung University, Tainan, Taiwan, October 15–16, pp. 1–18.

NSF 00-26, March 2000, *Program Solicitation, Exploratory Research on Model-Based Simulation, Directory for Engineering Division of Civil and Mechanical System*, National Science Foundation, Washington, DC.

Sotelino, E. D., White, D. W., and Chen, W. F., 1992, Domain-specific object-oriented environment for parallel computing, steel structures, *Journal of Singapore Structural Steel Society*, 3, 1, 47–60.

Sotelino, E. D., Chen, W. F., and White, D. W., 1998, Future challenges for structural engineering simulation, *Proceedings of the Sixth World Congress on Computational Mechanics in Conjunction with the Second*, Buenos Aires, Argentina, June 29–July 2.

Index